Edwin Moses Hale

The Medical, Surgical and Hygenic Treatment of Diseases of Women

Edwin Moses Hale

The Medical, Surgical and Hygenic Treatment of Diseases of Women

ISBN/EAN: 9783337811921

Printed in Europe, USA, Canada, Australia, Japan

Cover: Foto ©berggeist007 / pixelio.de

More available books at **www.hansebooks.com**

THE

MEDICAL, SURGICAL, AND HYGIENIC TREATMENT

OF

Diseases of Women,

ESPECIALLY THOSE CAUSING

Sterility,

THE DISORDERS AND ACCIDENTS OF PREGNANCY,

AND PAINFUL AND DIFFICULT LABOR.

Second Edition—Enlarged; with 55 Illustrations.

By EDWIN M. HALE, M.D.,

PROFESSOR OF MATERIA MEDICA AND THERAPEUTICS IN THE CHICAGO HOMŒOPATHIC COLLEGE; AUTHOR OF THE "MATERIA MEDICA, SYMPTOMATOLOGY AND THERAPEUTICS OF NEW REMEDIES;" OF "LECTURES ON DISEASES OF THE HEART;" MEMBER OF THE AMERICAN INSTITUTE OF HOMŒOPATHY; OF THE WESTERN ACADEMY OF HOMŒOPATHY;

HONORARY MEMBER OF THE HOMŒOPATHIC STATE MEDICAL SOCIETIES OF ILLINOIS, INDIANA, MICHIGAN, NEW YORK, MASSACHUSETTS, IOWA, WISCONSIN, AND OF THE HOMŒOPATHIC MEDICAL SOCIETY OF THE UNITED STATES OF COLOMBIA, SOUTH AMERICA; CORRESPONDING MEMBER OF THE MASSACHUSETTS SURGICAL AND GYNÆCOLOGICAL SOCIETY, ETC., ETC.

BOERICKE & TAFEL:
NEW YORK: PHILADELPHIA:
145 GRAND STREET.   635 ARCH STREET.
1880.

TO

SAMUEL LILIENTHAL, M.D.,

PROFESSOR OF CLINICAL AND PSYCHOLOGICAL MEDICINE IN THE NEW YORK HOMŒOPATHIC MEDICAL COLLEGE, IN TOKEN OF ADMIRATION OF HIS PROFESSIONAL HONESTY, LIBERALITY, AND EXTENSIVE LEARNING; OF GRATITUDE FOR NUMEROUS FAVORS, UNIFORM COURTESY, AND UNINTERRUPTED FRIENDLY SYMPATHY;

This Work

IS INSCRIBED BY HIS FRIEND,

THE AUTHOR.

PREFACE TO FIRST EDITION.

THIS work contains some of the results of the observation, study, and experience of a quarter of a century of active practice.

It is not intended to take the place of such works as Jahr, Leadam, Guernsey, Ruddock, Richardson, Marsden or Ludlam.

It occupies altogether another sphere in our literature and instead of superseding the above authorities, will serve to supplement them. I have purposely omitted the minute history, etiology, diagnosis and symptoms of many of the diseases treated of, referring the student to the larger systematic text-books, and the practical physician to his own knowledge and experience in relation thereto.

The time has arrived when Homœopathy must occupy a broader and more advanced position in the therapeutics relating to Gynæcology and Obstetrics, or its usefulness as a healing Art will remain circumscribed and contracted.

The Law of Cure must not be hampered by such rules as "the minimum dose, the single remedy, and the rejection of local treatment." The physician who practices according to this rule is only a *half*-homœopathist. He avails himself of but half the capabilities of our Law of Cure, and his sphere of usefulness is limited accordingly. There are conditions where material doses are as homœopathic as an immaterial attenuation. There are cases when medicines locally applied are as strictly homœopathic to the disease as when taken into the circulation by the mouth or stomach.

The Law of Cure, enunciated by Hahnemann, is universal and all-embracing. Palliatives may be useful as *aids* to a cure—in the same manner as surgical, mechanical and chemical measures—but no *cure* can be made by medicine, that is not a homœopathic cure.

The dose, the method of administration, and the mode of application of the remedy are all of minor importance, if the drug is selected under the law of *similia*.

In the following work I have not, except in rare instances, advised any remedy that is not homœopathic, either constitutionally or locally, or by virtue of its primary or secondary action. I yield to no one in my adherence to the law of *similia*, or my devotion to the cause of the homœopathic healing art, but I will not be bound down to old dogmas, or antiquated therapeutic notions. I believe the "sole duty of the physician is to cure the sick." Hahnemann taught us this duty.

To those who oppose all progress in therapeutics, outside of their restricted views, I would commend the following noble words of Hahnemann, written when he had been chastened by age and purified by adversity, and during the last years of his eventful life, when at last peace and happiness surrounded him:

"No reasonable physician can be satisfied with practicing within the limits of antiquated rules, derived from speculative theories, instead of pure experiments. His object is to cure the sick, and the innumerable powers of nature, WITHOUT EXCLUDING ANY, have been assigned to him to effect that process of regeneration. And to carry out such noble ends he ought to have a right to dispose of the curative powers of nature, for the preservation or partial regeneration of the body is a more noble deed than any of the boasted deeds of history; but he ought likewise to be permitted to employ those powers in THAT FORM AND QUANTITY which seem to him the most adequate and the most conformable to experience and wisdom; in this respect no restriction ought to be imposed upon him as a free and independent man, provided he is endowed with the necessary power and knowledge to preserve human life, and with that delicate conscience which every one whom God has appointed a guardian of human life should possess."

E. M. HALE.

CHICAGO, August, 1878.

PREFACE TO SECOND EDITION.

THIS second edition includes nearly all the special diseases of women.

The first edition of this work met with such rapid sale, and was received with such favor by the profession, that my publishers with that liberality which characterizes them, suggested that this edition be greatly enlarged and more profusely illustrated.

I have therefore tried to bring the book up to the most recent advances in Gynæcological Art, by adding the researches of the most eminent authorities in this department, as taught by their latest works and contributions. To this I have also added the experience of my colleagues, and my own.

The future of our School, in the art of treating the many diseases peculiar to women, will be marked by unexampled success, if we are not hampered by theories which tend to prevent progress. It is well for Homœopathy, that the majority are not content to discard all mechanical aids, all local treatment, and all palliatives, for such a practice would be the grave of our School, especially of its Gynæcology.

No system of practice equals ours in the possession of remedial agents wherewith to combat the disorders of women. But to be supremely successful with them we must be free from all dogmas relative to dose, and all blind reliance on mere symptomatology. We must (1) affiliate them to the organ or tissue diseased; (2) to the pathological condition of such organ or tissue; and (3) prescribe them, guided intelligently by characteristic symptoms and conditions—in such quantity as seems suited to each individual case.

Let it never again be said to our disparagement, that we allow displacements or local lesions to go on unarrested, because we reject simple and harmless appliances, or neglect mild topical applications.

A bungling or ignorant physician may injure a patient with an instrument or medicine, which, in the hands of an adept in his art, will result in prompt and lasting benefit.

E. M. HALE.

CHICAGO, August, 1879.

CONTENTS.

PART I.

STERILITY.

	PAGE
PREFACE TO FIRST EDITION, ...	v
PREFACE TO SECOND EDITION, ...	ix
INTRODUCTION: Discussion of the Facts and Theories of Ovulation and its Relation to Menstruation and Conception, ...	17–52

CHAPTER I.

ENUMERATION OF THE CAUSES OF STERILITY,   53–55

CHAPTER II.

STATISTICS OF STERILITY,   56–59

CHAPTER III.

CONSTITUTIONAL CAUSES AND THEIR TREATMENT: Chlorosis—Scrofula—Syphilis—Mercurialization—Obesity—Twin-Birth—Prostitution—Inordinate Sexual Intercourse—Changes of Climate—Mineral Waters—Improper Diet, Exercise, and Clothing, . . 60–66

CHAPTER IV.

PSYCHICAL CAUSES: Incompatibility—Frigidity—Erotism,   67–68

CHAPTER V.

OVARIAN CAUSES: Abscence of Ovaries—Imperfect Development—Atrophy—Inflammation—Degeneration—Tumors—Dropsy, . 69–71

CHAPTER VI.

UTERINE CAUSES: Stricture of Fallopian Tubes—Obstructions—Inflammation—Abscence or Defect of Uterus—Atresia of Uterus—Imperfectly Developed Uterus—Atrophy of Uterus—Senile Atrophy—Obstructions of the Os and Canal of the Cervix—Simpson's operation and Sims' operation—Superficial Trachelotomy—Stricture of the Canal—Displacements of the Uterus—Anteversion and Flexion Retroversion and Flexion—Lateroflexion—Prolapsus—Elevation

xii   CONTENTS.

—Inversion—Tumors—Polypi—Fibroids—Endometritis—Endocervicitis—Abrasion—Erosion—Ulceration—Leucorrhœa—Dysmenorrhœa—Amenorrhœa—Menstrual Irregularities—Areolar Hyperplasia—Abnormal shape of Cervix—Laceration of Cervix Uteri, ...   72-204

CHAPTER VII.

VAGINAL CAUSES: Atresia—Abscence of—Non-retaining—Imperforate Hymen—Vaginismus—Leucorrhœa—Disease of Clitoris, . .   204-221

CHAPTER VIII.

RENAL CAUSES: Diabetes Mellitus—Diabetes Insipidus, .   222

CHAPTER IX.

VESICAL CAUSES: Irritable Bladder—Catarrh of Bladder, . .   223-226

CHAPTER X.

URETHRAL CAUSES: Irritable Urethra—Chronic Catarrh Urethritis—Caruncles of Urethra—Granular Erosion of Urethra—Fissure of Urethra—Vesico-vaginal Fistula—Vascular Tumor of Meatus, .   227-230

CHAPTER XI.

RECTAL CAUSES: Hæmorrhoids—Prolapsus Recti—Fissure of Anus—Fissure of Rectum—Ulcer of Rectum, ...   231-234

CHAPTER XII.

THE MECHANISM OF CONCEPTION: The Nature and Properties of Semen—Spermatozoa—Examination of Vaginal and Cervical Mucus for Spermatozoa—Hale's Speculum—The Passage of Spermatozoa to the Ovule—The Successful Conditions of such Passages, .   235-247

CHAPTER XIII.

HYGIENIC AND OTHER ERRORS CAUSING STERILITY: Improprieties of Dress—Imprudence during Menstruation—Excessive Venery and Orgasm—Improper Time for Coition—Coition during Treatment of Uterine Disease—Abortive Insemination—Improper Position during Coition—Improper Conduct after Coition—Means Used to Prevent Conception—Abortion as a Cause of Sterility, . .   248-261

CHAPTER XIV.

GENERAL THERAPEUTICS OF STERILITY: Indications for the Use of Various Medicines—Miscellaneous Therapeutic Agents—Galvanism—Carbonic Acid Gas—Gymnastics—Hydropathy—Mammary Irritation—Injection of Semen into the Uterus, ...   262-278

PART II.

DISORDERS OF PREGNANCY AND DYSTOCIA.

	PAGE
PREFACE,	281

CHAPTER I.

GENERAL DISORDERS OF PREGNANCY, REMEDIES FOR: Preventive Treatment of Dystocia, ... 283–301

CHAPTER II.

SPECIAL DISORDERS AND ACCIDENTS OF PREGNANCY: Vomiting—Retroflexion of the Uterus—Cases Illustrating—Use of Pessaries in—Albuminuria—Treatment of—Urænic Poisoning—Convulsions—Constipation—Treatment of, ... 302–333

CHAPTER III.

MEDICATION OF FŒTUS IN UTERO AS A MEANS OF PREVENTING DYSTOCIA: Dystocia Due to the Fœtus—Unnatural Ossification of the Skull—Hydrocephalus of the Fœtus—Tumors of—Montrosities—Hardness of the Bones of—Fruit Diet for Dystocia, . . . 334–347

CHAPTER IV.

THE IMMEDIATE TREATMENT OF FUNCTIONAL DYSTOCIA: Extreme Slowness of Labor—Rigidity of Cervix—Spasmodic Contraction of the Cervix—Treatment of—Second Stage of Labor—Feebleness of Contractions—Excessive Distention of the Uterine Walls—Sudden Cessation of Pains—Cramps in the Legs—Irregularity of the Pains—Position during Labor—Use of the Hale's Small Forceps during Difficult Labor, ... 348–375

LIST OF ILLUSTRATIONS.

FIG.		PAGE
	Frontispiece, (3 figures) illustrating the normal position of uterus, anteflexion and retroflexion).	
1.	Fitch's Measuring Sound,	75
2.	Simpson's Metrotome,	83
3.	Illustrating Greenhalgh's Operation,	84
4.	Section of Uterus Internum,	86
5.	Section of Uterus Internum,	86
6.	Discission of the Cervix,	91
7.	Peaslee's Metrotome,	101
8.	Normal Cervical Canal,	103
9.	Normal Cervical Canal, Modified by Peaslee's Method,	103
10.	Uterine Cavity after Sims's Operation,	104
11.	Uterine Cavity after Simpson's Operation,	104
12.	White's Hysterotome,	104
13.	Hale's Expanding Speculum,	105
14.	Nelson's Tenaculum,	105
15.	Sims's Cervical Dilator,	106
16.	Hunter's Dilator,	108
17.	Chamber's Split-stem Pessary,	111
18.	Sargent's Galvanic Stem Pessary,	112
19.	Thomas's Anteversion Pessary (old),	113
20.	Thomas's New Anteversion Pessary (shut),	113
21.	Thomas's New Anteversion Pessary (open),	114
22.	Exploring and Indicating Sound,	116
23.	Exploring and Indicating Sound, manner of using,	117
24.	Exploring and Indicating Sound, manner of using,	117
25.	Thomas's Retroversion Pessary,	119
26.	Albert Smith Pessary,	119
27.	Studley's Straight Stem Pessary,	120
28.	Beebe's Anteversion Pessary,	135
29.	Beebe's Anteversion Pessary *in situ*,	137

FIG.		PAGE
30.	Beebe's Anteversion Pessary *in situ*,	138
31.	Simpsons' Sound,	141
32.	Sims's Sound,	141
33.	Sims's Elevator,	141
34.	Gardner's Elevator,	141
35.	Elastic Ring Pessary,	144
36.	Molesworth's Double Canula and Bulb Syringe,	156
37.	Nott's Double Canula Catheters,	156
38.	Wylies Cervical Protector,	157
39.	Bozeman's Dressing Forceps,	163
40.	Siemen's Scoop,	164
41.	Thomas's Serrated Curette,	164
42.	Sims's Cervical Applicator,	168
43.	Hale's Cervical and Intrauterine Applicator,	168
44.	Sims's Porte-tampon,	174
45.	Chapman's Scarificator,	184
46.	Buttle's Spear Scarificator,	184
47.	Skeene's Hysterotome,	195
48.	Peaslee's Uterotome,	196
49.	Conical Cervix,	197
50.	Lacerations of the Cervix Uteri,	202
51.	Hale's Small Expanding Speculum,	212
52.	Spermatozoa,	236
53.	Hale's Large Expanding Speculum,	242
54.	Hale's Forceps for Rotation and Extension,	370
55.	Method of using Hale's Forceps,	374

ON STERILITY.

INTRODUCTION.

IN the treatment of this subject I propose to adopt the generally accepted meaning of the term, namely: the want of the *aptitude* for being impregnated.

This definition does not restrict barrenness to those cases where, from some cause, ovulation does not take place; for that function may be perfectly performed, and yet, from some abnormal condition of the organs of generation, the ovule cannot become impregnated.

Impotence in the male cannot be considered as a cause of sterility, and the consideration of this condition does not fall within the province of this treatise.

Sterility may be *congenital* or *acquired;* or it may exist for years, and a spontaneous cure result from the action of unknown causes.

History relates many instances illustrating these conditions. Anne of Austria had been married to Louis XIV for fifteen years before she had a child. Another Anne of Austria, wife of Louis XIII, was sterile for twenty-two years. Catalina de Medicis, wife of Henry the Second, was unfruitful during the first ten years of her marriage, after which time she became so prolific that she had ten successive children. Many similar cases are reported in the various works on obstetrics.

There is still another class of sterile women, who become so after having one or two children, and are afterwards barren, even when no cause is apparent in the condition of the organs of reproduction.

For a proper understanding of the subject, and a full appreciation of the various causes of sterility, the latest views on the physiology of generation should be kept in mind.

It is generally believed that conception consists in the contact of the fertilizing semen of the male with the ovum, *somewhere* in the Fallopian tubes, or the cavity of the uterus.

It is taught by some authors, however, that the ovum may or does become impregnated in the ovary. It is established that the spermatozoa after being deposited in the vagina, even at its entrance, find their way to the cervical canal, which they enter, and thence proceed until they meet the ovum, which they enter. This is impregnation. The ovum passes into the cavity of the uterus, and if impregnated then or during its progress to the cavity, becomes attached to the interior surface of that organ.

But the subject of *ovulation*, and its connection with menstruation and *coition*, is one which more concerns us, for unless *ovulation* and *coition* coincide, even in a perfectly normal condition of all the organs of both sexes, *impregnation* will not result.

It was my intention to collect and place before the student and physicians the very latest observations and facts relating to ovulation; but I fortunately found a paper written by Dr. A. R. Jackson, of Chicago, published in the *American Journal of Obstetrics*, October, 1876, which contains all observations of importance up to the present date.

By his permission I give his paper entire. After carefully reading it this deduction will be evident to any practical mind, namely, that in some cases of sterility the parties must be informed of the main facts of the process of ovulation, in order that they may try to make the period of *ovulation* and *coition* coincident.

While I admit that probably no exact information as to the time of ovulation can be given by the physician, it is probable that careful experiment may enable the husband and wife to ascertain its occurrence.

Dr. Jackson says: "It is my purpose, in this paper, to examine the evidence upon which is based the ovular or ovulation theory of menstruation.

"This subject has received a large share of attention during the past few years, and a great discrepancy of opinion exists in regard to many points connected with it. I do not expect to reconcile these discordant views; indeed, in the present state of our knowledge, it would perhaps be impossible to do so. Our great need in this, as in many other problems of a physiological character, is an increased number of well-observed facts; those which we have thus far are too few in number, and, apparently, too contradictory to warrant a definite and entirely satisfactory conclusion.

"The progress of scientific knowledge is greatly retarded by the admission of what may be termed, paradoxically, false facts; that is to say, facts which by representing only partially the truth, lead to false results. Conclusions founded upon such premises must almost necessarily be erroneous. Truth in science is rarely found wholly unmixed with error, and, in order that we may rightly appreciate the former, we must properly estimate the latter also. Like the diamond, whose facets reflect a differently colored ray according to the angle from which they are seen, so may a scientific truth present a different aspect to those who behold it from different points of observation. While to one person the gem appears green, to another it is red, and to a third yellow. He only who sees it from all directions can know the whole truth. Thus it is that imperfectly observed, partial or perverted facts result in the formation of a false theory; and a false theory, once adopted, has a most injurious influence. He who is governed by it sees everything through a false medium. As observed by Paris, 'He who is guided by preconceived opinions, may be compared to a spectator who views the surrounding objects through colored glasses, each assuming a tinge similar to that of the glass employed.' The advocates of the ovulation theory are, it seems to me, somewhat in this position. Many facts have been observed which give apparent support to their opinions, and on these they have been content to rest, overlooking, or, at least underestimating other facts, equally well known, which strongly militate against those opinions.

"The ovulation theory of menstruation implies the following essential propositions:

"1. At regular periods, of about twenty-eight days, in the human female, a matured ovule is discharged from the ovary, passes into the Fallopian tube, and is transmitted to the uterus.

"2. Coincident with and dependent upon the maturing and bursting of the Graafian vesicle and the extrusion of the ovule, certain changes occur in the mucous membrane of the body of the uterus, which result in a sanguineous change from that organ.

"In support of these propositions, evidence consisting of certain facts and analogies has been adduced, as follows: (a) Observations made on the bodies of women who have died during or soon after the menstrual period have revealed the presence in one or other ovary of a ruptured Graafian vesicle, and its cavity filled with a blood-coagulum, or its remains, a corpus luteum, in various stages of development or decadence; (b) Physiologically, the period of menstruation in woman corresponds with the rut or œstrus of other mammalia, when, it is well known, ova are discharged from the ovaria; (c) The artificial removal of the ovaries causes an immediate cessation of the menstrual function. I purpose considering, *seriatim*, these propositions, together with the facts which have been advanced for their support.

"1. AT REGULAR MONTHLY PERIODS IN THE HUMAN FEMALE, AND COINCIDENT WITH THE MONTHLY FLOW, AN OVULE IS DISCHARGED FROM THE OVARY, IS RECEIVED INTO THE FALLOPIAN TUBE, AND BY IT TRANSMITTED TO THE UTERUS.

"The minor proposition, namely, that the matured ovule passes from the ovary to the uterus through the Fallopian tube is admitted on all hands, and, not being in dispute, need not detain us. The essence of the controversy centres in the alleged periodicity of this process, and of its time relations with the menstrual discharge.

"The Graafian vesicles, from the time of their description in 1673, by De Graaf, down to the year 1827, were thought to be the actual ova of mammalia. It was not until the last-named period that Baer discovered the true ovule and the relations it bore to its containing vesicle. However, as early as 1672, Kerkringius* advanced the idea that the ova were discharged at the time of menstruation, but it does not seem to have been founded upon any observations. The first writer who gives positive evidence upon the subject is Sir Everard Home, who noticed the ruptured follicle during menstruation, although its import was not then understood. In 1821, Dr. Power clearly enunciated the doctrine of the periodical ripening of the follicle at the menstrual period; and the discovery six years later by Baer, already alluded to, that this was only the enveloping structure of the ovule and not the ovule itself, made the rupture an intelligible fact; and so we may regard this as the real birth of the ovulation theory. In 1831, Negrier,† working independently, showed by anatomical preparations that the periodical discharge of menstruation was the consequence of an internal hidden function—ovulation. Fresh proofs were brought forward by Gendrin, Paterson, Barry, Raciborski, Bischoff, Pouchett, and others, all tending to show that ovulation and menstruation are simultaneous and necessarily connected one with the other; and the doctrine was so beautiful and reasonable, and seemed so well sustained by the evidence adduced, that we cannot wonder at the fact that it was generally received and adopted by physiologists. Still, there have always been some who were not convinced of its correctness, and who regarded the proofs alluded to as insufficient and inconclusive; who, in the language of Mr. Kesteven,‡ looked upon the doctrine ' as a plausible and ingenious theory, wanting, however, in the true elements of an inductive theory; in short, an example of the *post ergo propter* line of argument.'

" In examining the cases which have been cited in support of the ovulation theory, one cannot fail to be struck with the complacency with which conclusions are frequently drawn from irrelevant, or, in some instances, even adverse facts.

" An example of this is to be found in a review of Bischoff's work on *Human Ovulation*, in the *American Journal of the Medical Sciences*, vol. 28, p. 137.

---

* Tyler Smith, Lectures on Obstetrics, third edition, p. 80.
† Recueil de Faits pour servir à l'Histoire des Ovaires. Angers, 1858.
‡ Lond. Med. Gazette, 1849.

"These cases of Bischoff have always been regarded with especial favor and as of great value by the advocates of the ovular theory; and, as they are frequently alluded to, I feel constrained to present a very brief synopsis of them.

"The observations* were thirteen in number. Of these, the time of the menstrual period was know in only ten; the remaining three have therefore no value so far as this inquiry is concerned. Of the ten, three died during menstruation, and in each of these there was found a ruptured follicle. A fourth died two days after menstruation: the right ovary contained a pretty large projecting follicle, which was *still closed*. Both ovaries contained small corpora lutea. In a fifth case, the menstrual period had just passed; the left ovary contained a very distinct corpus luteum, and the right ovary a ruptured Graafian vesicle filled with fresh blood. Number six died seven days after menstruation: in the right ovary was a recent corpus luteum. In the seventh, death occurred ten days after menstruation: the right ovary contained a very large Graafian follicle *unopened*. Number eight died ten days after menstruation: the right ovary contained a *recently ruptured Graafian follicle and a fresh corpus luteum*. In number nine, menstruation had occurred eighteen days before death: the right ovary contained a very large corpus luteum. Lastly, number ten died four weeks after menstruation; the right ovary contained a ruptured Graafian follicle.

"After detailing these cases, the reviewer says: 'The results here obtained show that in the human female, at each menstrual period, a Graafian follicle is ripened, swells and, usually bursts, discharging an egg,† and forming a corpus luteum.' Now, I submit to any candid inquirer that the cases cited do not show these things. Indeed, so far as they prove anything, it is that there is not even an approximate correspondence between the rupture of the follicle and the menstrual period. In two of the cases menstruation had occurred without any

---

* Beiträge zur Lehre von der Menstruation und Befruchtung, 1853.

† Notwithstanding the most diligent search, Bischoff was unable, in a single instance, to discover the ovule.

such rupture at all. In Cases 5 and 8 the ovaries are described as containing a fresh corpus luteum *and* a recently ruptured follicle; yet, in the one case, menstruation had 'just passed,' and in the other had ceased ten days before death. Inasmuch as a corpus luteum is an older formation than a recently ruptured follicle, we should naturally refer the latter to the last menstrual period. But to what period or periods do the *fresh corpora lutea* belong? Dalton* says that the corpus luteum of menstruation 'reaches its greatest development about three weeks after ovulation, and from this time rapidly disappears, a small cicatrice only remaining.' This being the case, can we refer these 'fresh' corpora lutea to a menstrual period, in one case thirty, and the other thirty-eight days past? If we can do so,—if a recent corpus luteum signifies one which may be associated with a menstruation which occurred four to six weeks before—then what shall we say of Case 6, where the presence of such a one is connected with a period *seven days* past?

"Barnes,† too, with a similar disregard of consistency, after stating that the preparations of Coste, preserved in the College of France, prove that the ripening of a Graafian follicle always coincides with the turgescence of the genital organs, and, according as the circumstances are more or less favorable, bursts at the commencement, towards the end, or at any time during the menstrual discharge—proceeds to state what these preparations are in detail, thus: 'In a woman who died on the first day of the appearance of the menses, the ovarian vesicle was manifestly ruptured. In another, who died four or five days after the cessation of the menses, the right ovary presented a vesicle still intact, but so distended that the slightest pressure made it burst. Lastly, in a young virgin, who died fifteen days after menstruation, there was no recent trace of a yellow body, and it could not be doubted that the Graafian vesicle had been arrested in its development.' Surely, these cases, taken together, so far from proving that a Graafian follicle bursts at the menstrual period, show that menstruation occurs,

---

* Prize Essay "On the Corpus Luteum of Menstruation and Pregnancy."
† Diseases of Women, p. 147.

in two-thirds of the cases, without such rupture, and in one-third without even a maturation of a follicle. For in one only of the three was there actually found a ruptured follicle; in one, menstruation had occurred and ceased several days before without any rupture, the follicle being burst by external force post-mortem, and in the third there had not been even the ripening of one!

"Such a course as is indicated in the foregoing instances is more reprehensible than reasoning without facts; it is reasoning against them. Yet these are only samples of the sort of argumentation which is frequently found in connection with this subject.

"(A.) MENSTRUATION MAY OCCUR WITHOUT ACCOMPANYING OVULATION.

"At first glance it would seem that we ought to accept the discovery of a rent follicle filled with blood, in persons who have died during menstruation, as a proof that the flow is, if not the result of, at least coincident with, ovulation. But such evidence is not at all conclusive, and may be erroneous, for Ritchie* has 'repeatedly seen the opening of a discharged vesicle to be still patent, and sometimes the vesicle to be filled with a florid blood-clot in the third and fourth month of pregnancy; and, in one case, he found the corpus luteum of a woman in the ninth month to communicate with the surface by a distinct foramen.'

"We must bear in mind, in the consideration of this subject, that Graafian vesicles are maturing and rupturing, and corpora lutea are forming and disappearing continually; hence it should be expected that, in a woman dying at almost any time, some of these conditions would be found in the ovary. And when we further consider that the menstrual periods occupy from one-sixth to one-fourth of a woman's lifetime for thirty years, we may equally expect to find such ovarian changes at these periods also. So that it seems strange, indeed, that

---

* Tilt, Uterine and Ovarian Inflammation, p. 66.

persons who are satisfied of the correctness of the ovular theory because occasionally a ruptured follicle or a corpus luteum is found coincident with a menstrual period, should ever have a lack of evidence. And yet abundant as such evidence unquestionably is, it does sometimes fail; for menstruation frequently occurs without any such contemporaneous change in the ovary. Many instances of this character have been recorded, but it is only necessary to call attention to a few of them.

"Dr. W. W. Gerhard* presented to the College of Physicians of Philadelphia the uterus and appendages of a multipara, twenty-five years old, who died of apoplexy during a menstrual period. 'At several points on the surface of the ovary there were minute dot-like orifices, each one corresponding to a Graafian follicle. Two of these being examined under the microscope, were found to present a few granular nucleated cells floating in a homogeneous liquid.' Although this woman had been the subject of menorrhagia, the discharge latterly returning profusely every two weeks, there was no evidence of any recent ripening or rupture of a follicle. Again, Dr. Stedman† has reported the case of a married woman, forty-five years of age, who died of some pulmonary affection. Menstruation was regular nearly to the time of her death, and yet on examination there was found no trace of the left ovary, but in its place a thin and simple serous cyst nearly two and a half inches in diameter; while on the other side there was a collection of cysts, forming a mass twice the size of an English walnut, upon the surface of which were spread out the thin flattened atrophied remains of the ovary.

"Futhermore, it is the experience of ovariotomists, that in many cases in which both ovaries have been removed, these organs have been found so thoroughly diseased as to preclude the idea that they could possibly have performed their function of ovulation normally, if at all, and yet the regularity of menstruation has suffered no interruption.

"Some of the reported cases of hernia of the ovaries throw

---

* Amer. Jour. Med. Science, vol. xxxvi, p. 410.
† Ib., vol. xxiv, p. 83.

valuable light upon this question. For example, Dr. Oldham* presented one to the Royal Society, the subject of which was a tall, well-formed woman, nineteen years of age, in whom both ovaries had descended through the inguinal canals, and occupied positions in the upper part of the labia majora. The mammæ and external genital organs were well developed, but neither uterus nor vagina could be detected. The left ovary was in a quiescent state, and had never been the seat of pain or swelling. She was under Dr. Oldham's observation six years, during which time he had frequent opportunities of seeing her. For the first three years the right ovary was exclusively enlarged, the intervals varying from three weeks to three months. For the last two years the left ovary was most frequently affected, the right remaining quiescent. Occasionally both were tumid, but one always more so than the other. The swelling sometimes occurred suddenly, although usually it was gradual, the volume of the organ increasing slowly for four days, remaining stationary for three days, and then slowly declining, the whole process lasting ten or twelve days. During this period the organ was tender when pressed, but was otherwise not painful, and did not interrupt the patient's ordinary duties. There were no manifest sympathies excited in the mammary glands or other organs; and there was no vicarious flux, either of blood or other secretion. The ovary alone seemed engaged in the act. It was supposed, reasonably, that these periods of enlargement were those of ovulation, and I beg to call attention to the fact that they were quite irregular in their occurrence, the intervals varying from three weeks to as many months.

"Dr. Alfred Meadows,† also, mentions a case of similar character. The patient was a single woman, twenty-three years of age, who began to menstruate at fifteen, and continued doing so at regular intervals, with some pain, down to the age of twenty, when, after stooping, a swelling suddenly appeared in the right inguinal region, caused, as was subsequently learned,

---

* Ib., vol. xxxv, p. 284.
† Amer. Jour. Obstetrics, etc., vol. vi, p. 231.

by the prolapsed ovary. At the menstrual period following this she suffered violent pain of a character different from any she had experienced before; it preceded the hemorrhage; at the same time the tumor was much increased in size. From that time on she suffered in a similar way, sometimes more acutely, so that at every monthly period she was obliged to lie in bed for a week or more. Sometimes the tumor would swell up to the size of 'two fists,' and be exquisitely tender to the touch. She had no suffering during the intermenstrual periods. To what does this history point? Are we not prepared to accept it as an evidence of the truth of the ovular theory? Prior to the occurrence of each monthly flow, for a period of three years, the ovary enlarges, becomes the seat of pain, and the swelling does not subside until after the cessation of the discharge. What more ought we to require?

"Mark the sequel. At the suggestion of Dr. Meadows the tumor was removed. It was not contained in any cyst or sac, and was readily separated from its fatty and cellular attachments. The upper portion or pedicle, which went through the abdominal ring, was found distended with fluid. This was punctured, and about an ounce of the contents let out. The pedicle was then tied, and the tumor removed. 'The tumor, which measured about two inches in diameter, proved, on section, to be the right ovary. It had, however, undergone remarkable structural change. Instead of presenting the usual dense compact appearance, it contained throughout numerous irregularly shaped spaces, varying in size from a pin's head to a quarter or even half an inch, and all were filled with the same kind of fluid as flowed from the pedicle. These cells appeared to communicate with one another, and the whole organ to be infiltrated, as it were, with the fluid in question. There were no proper Graafian vesicles to be seen.' No Graafian vesicles,—no ovules—no ovulation, to account for the great increase in the size of the ovary preceding and during the catamenial period. What is the plain inference? Is it not that the swelling of the ovary was caused by the pelvic congestion attendant upon the menstrual period? And yet Dr. Meadows has introduced the account of this case in an argument affirming, among other things, the dominating influence of the ovaria, and the fact of ovulation producing the menstrual flow.

"This last-mentioned case of ovarian hernia, as also one reported by Dr. McCluer,* and another which I am informed† has been published by Dr. Joseph English, of Vienna, shows that the cystic degeneration of the prolapsed ovary, even when its essential vesicular structure is wholly destroyed, does not prevent the organ from enlarging and becoming painful during the menstrual epoch. In all of these cases the ovary, which had swollen month after month at regular periods corresponding with the menstrual flow, was found to be so diseased as to leave no vestige of Graafian vesicles, thus proving ovulation to be impossible.

"Dr. Tilt says: 'In three cases in which Dr. Ashwell had opportunities of examining the ovaria of women who died during the flow of the catamenia, there were no signs of the rupture of the Graafian vesicle and the escape of ovules. In one of these cases the woman had menstruated regularly for several years, and yet the ovaria were perfectly smooth; there was neither rent nor cicatrix marking the site of either a present or former maturation and escape of a Graafian vesicle.' Ritchie‡ also reports five examples of menstruation which were not accompanied, and could not have been caused by ovulation. In one of these the woman died ten days after menstruation. The ovaries were filled with vesicles, but neither of them presented either a puncture or cicatrix. In another, death occurred thirteen days after menstruation. Here, too, the ovaries contained numerous vesicles, one as large as a garden pea, but in neither of them was puncture or cicatrix. In a third menstruation had occurred a week before death, but there was neither scar nor opening on the surface of either ovary. Again, in another case where death occurred a fortnight after menstruation, neither ovary presented any sign of recent rupture.

---

* Amer. Jour. Obs., vol. vi, p. 613.
† Dr. Paul F. Mundé, private letter.
‡ Ovarian Phys. and Path., London, 1865.

"Dr. John Williams* has published a series of cases bearing upon the temporal relations of the discharge of ova with the menstrual flow. He believes that the ova are discharged, usually, before the appearance of the catamenial flux, and details observations made by him upon sixteen cases. In several of these, where death occurred during the intermenstrual period, his conclusions are drawn from the condition of the corpus luteum; but inasmuch as the changes in these bodies do not always take place uniformly, and, as it is always difficult to determine the age of effused blood, the results founded upon these cannot be accepted as certainly accurate. Passing by these, therefore, I wish to invite attention to those of his cases in which death occurred during the menstrual period.

"(1.) Was a young woman who died on the fifth day of the flow. 'On the surface of the left ovary was a rough, brownish-colored, star-like cicatrix. On section there was seen under the cicatrix a corpus luteum dilated in the middle and narrow at both ends, nearly three-quarters of an inch in length and half an inch in width; its walls were in some parts of a pinkish, and in others of a yellowish color. In the centre was a partially decolorized clot.' (2.) Was a patient who died on the ninth day of typhoid fever and the fourth day of menstruation. One ovary contained a corpus luteum, similar to that in case No. 1. In both of these rupture of the follicle had taken place evidently several days before. (3.) Woman had undergone operation for fistula in ano, and died five days after the appearance of menstruation; one ovary contained a follicle five-eighths of an inch by one-third of an inch, in which was found a bright-red, fresh, loose clot, and its walls were thin and smooth. *No rupture had taken place.* (4.) Patient with fibroid tumor of the uterus; died on the third or fourth day of menstruation. Left ovary contained a follicle nearly an inch in length, in which was found a soft, dark-colored clot, which appeared to be several days old; *follicle had not ruptured.* (5.) Patient died when the menstrual flow

---

* Obstet. Jour. Great Britain and Ireland, vol. iii, p. 620.

had almost ceased. *There was no rupture in either ovary*, but the right ovary contained a Graafian follicle about the size of a small pea. (G.) Young suicide; died three days after cessation of the flow. *There was no recent rupture in either ovary;* the left contained a follicle similar to the preceding.

"Of the foregoing six cases, in only two did a ruptured Graafian vesicle even seem to correspond with a menstrual period; and in two of the cases, the follicles most advanced were so immature that Dr. Williams expresses the opinion that they would probably have ripened by the next return of the flow.

"Mr. Paget* has reported a case of a woman who was executed for some crime, and the post-mortem appearances tell very forcibly against the ovular theory. The woman had begun to menstruate twelve hours before her execution. 'The ovaries were of moderate size and presented numerous marks of cicatrices upon their surfaces. In the right ovary, three Graafian follicles projected slightly on the surface and looked healthy, containing clear serous fluid. A fourth was of a very large size and prominent. In the left ovary, one Graafian follicle was fully developed and prominent. We looked for ova in the contents of all these, but in vain. The surface of the ovaries was generally rather more than usually vascular, but there was no peculiarly vascular spot, nor any appearance of the recent rupture of a vesicle, or the discharge of an ovum. In the right ovary, near the surface, was a small cyst or cavity, containing what looked like a decolorized clot, and bounded by a thin layer of a bright yellow-ochre substance, and excellent example of a fibro-corpus luteum, of one or more months' date, certainly not more recent.'

"Dalton, in the essay on the Corpus Luteum already referred to, reports two cases, in one of which death occurred during the menstrual period and in the other at its termination. In neither of them had a follicle recently ruptured, although in the second there was one on the point of doing so.

"I have had two opportunities of examining the ovaries of

---

* Tilt's Uterine and Ovarian Inflammation, p. 64.

women who died at or near the menstrual period. One was the case of a healthy unmarried woman, twenty-eight years of age, who died from an overdose of morphia, taken accidentally. She had menstruated regularly, and a period had ceased four days before death. Both ovaries were normal in structure and size, the right being somewhat larger than the left. It contained several Graafian vesicles scattered throughout the stroma. Two of these were larger than the rest, one being about an eighth of an inch in diameter, and the other as large as a small currant. This latter was near the surface and caused a slight projection. It contained a clear serous fluid. The left ovary contained fewer vesicles, but had the indistinct remains of a corpus luteum not less, certainly, than four or five weeks old. The mucous membrane of the uterine body was pale and covered with a grayish-pink mucus. The other case was that of a young girl, fifteen years old, who died from the effects of a burn. She had commenced menstruating eighteen months before, but the function had been regularly performed only for about ten months. A period had ceased twelve days prior to death. Neither ovary contained corpora lutea, nor bore the marks of recent rupture. The largest vesicle, which was about a quarter of an inch in diameter, was found in the left ovary about a sixteenth of an inch from the surface.

"(B.) OVULATION MAY OCCUR WITHOUT ACCOMPANYING MENSTRUATION.

"I will next proceed to adduce evidence to show that ovulation certainly and frequently takes place without menstruation.

"Malpighi and Vallisneri long ago observed that fully developed Graafian vesicles are occasionally found in the fully grown fœtus. Ritchie,[*] also, has demonstrated by at least ten dissections that in the ovaries of newborn infants, and children as early as the sixth year, may be found highly vascular Graafian vesicles; and that at the age of fourteen, and prior

---

[*] Loc. cit., p. 62.

to menstruation, they are found as large as small raisins, filled with their usual transparent granular fluid; that menstruation is not essential, either as cause or effect of these conditions; that prior to menstruation, the vesicles are found at every other period of life, in continual progression towards the circumference of the ovaries, which they penetrate, discharging themselves through the peritoneal coat, thus proving that the catamenial flow is not an indispensable prerequisite to their rupture.

"So, likewise, the more recent researches of Grohe, Slavjansky, and Haussmann have shown that the growth of the Graafian vesicle is quite independent of the menstrual period; and the last-named authority,* whose observations were made upon eighty-four subjects, asserts that such early development of the follicles as was noticed by Ritchie takes place in about ten per cent. of all cases. Dr. Sinety† confirms these observations, and maintains that 'in the ovaries of the newly born, Graafian follicles are almost always visible to the naked eye; and they may at this time often be discovered as well developed as in the adult female, and constituting true cystic ovaries, in which are to be seen ovules whose origin is indubitable. In the ovaries of infants, there are often cicatrices and follicles in different stages of atrophy.' What, I would ask, causes the cicatrices? Is it aught but the rupture of the Graafian follicles which, as would seem from the foregoing, may take place at any period of infantile and adult life?

"Slavjansky,‡ who has devoted a great deal of time to researches on the physiology and pathology of the ovaries, thus summarizes the results obtained by him:

"1. The Graafian follicles develop themselves from the primordial follicles, and are growing towards maturity from the first month of birth to the fortieth year. 2. The larger number of follicles do not mature, do not rupture, do not discharge their contents, but pass over into a condition of atrophy which is analogous to the formation of the corpora lutea.

---

* Centralblatt, No. 2.
† Le Progès Médical.
‡ Allg. Med. Centr. Z., 54, 1874.

3. The development and ripening of the Graafian follicles do not take place periodically in a regular manner, and there is no connection between ovulation and menstruation. 4. Menstruation is a physiological phenomenon unconnected with the development and ripening of the Graafian follicle. 5. The rupture of the more or less ripe follicle is associated with congestion of the genital organs, and is, as yet, an unexplained matter.

"On the other hand, ovulation may continue after the menopause. Lawson Tait* says: 'The cessation of the menses at the climacteric, though it diminishes the activity of the cell-growth at once to a marked extent, never extinguishes it; for the development and extrusion of immature Graafian follicles ceases only with life itself. They are to be found of some size even fifteen or twenty years after the cessation of menstruation.'

"Of course, ovulation is the necessary condition of impregnation, and it is admitted by all writers that conception may occur in the absence of menstruation. Our literature contains many instances of girls who have conceived prior to the first appearance of the flow; of women who have become pregnant subsequent to the menopause, and during lactation before menstruation has reappeared. Dr. James Young,† Tanner,‡ Dubois,§ Tilt,‖ and, indeed, almost every obstetric author, mention cases of this character. Leishman speaks of a woman who married at twenty-seven, and who menstruated the first time two months after her eighth labor. Raciborski states that he has seen on the ovaries one or two cicatrices, although the subjects had never menstruated.

"An argument which has been frequently urged in support of the ovular theory, is the fact that conception is more likely to take place shortly after a menstrual period than at any other time.

---

* Hastings, Prize Essay of 1873, London, p. 4.
† Am. Jour. Med. Science, vol. ix, p. 568.
‡ Handbook of Pract. Obstetrics, p. 24.
§ Journal de Méd., 1850.
‖ Uterine and Ovarian Inflammation, p. 49.

"Dr. W. H. Studley,* alluding to this, considers it as admitting of a very different explanation, and not at all as proving the coincidence of ovulation and the menstrual flow. He says: 'My opinion in regard to the rationale of the fact is this: impregnation is more likely to be secured at this time because of the recent deluging with menstrual blood, by which the secretions, especially of the cervical canal, have been washed away, which secretions often prevent impregnation either by their chemical incompatibility with the vitalizing fluid, or by the mechanical obstruction in the form of the firm mucous plug so often found in the canal.'

"If the ovular theory were true, conception could take place only at or near a menstrual period; but there is abundant evidence to show that it may and does frequently occur at times quite remote from it. My own experience has furnished me with a number of instances where married women anxious to prevent an increase of family, have observed the 'physiological rule' of abstinence for a fortnight after a period, and who have found, to their chagrin, after a time, that their precaution had been unsuccessful.

"Dr. Oldham observes: 'I know of cases which I have carefully inquired into, where impregnation occurred at the respective times of ten, twelve, and twenty-one days after the menstrual period: and while, on one hand, I am quite ready to admit a *greater* disposition to impregnation shortly after a menstrual period, yet I know of no facts to disprove the opinion that the human female is susceptible of impregnation at any time between her monthly periods.' Hirsch,† likewise, has seen a case where impregnation took place twenty-two days after a normal menstrual period; and he observes that, 'as the Jewish women are obliged to abstain from intercourse five days before and seven days after menstruating, that race could not be so prolific as it is known to be if the ovular theory of menstruation is true.' Tilt,‡ also, mentions the case of a lady, aged forty-seven years, in whom menstruation had been irreg-

---
* Amer. Jour. Obstetrics, 1875, p. 487.
† Schmidt's Jahrbuch, 1853, No. 2.
‡ Change of Life, p. 69.

ular for two years, and who after a single coitus, seventeen days subsequent to a period, became pregnant.

"An attempt has been made to explain these and similar cases, by supposing that the spermatozoa, on the one hand, and the ovule, on the other, may retain their vitality in the generative passages for a sufficiently long time to permit the occurrence of impregnation under the circumstances named. But the facts bearing upon the subject, so far as known, do not justify such an explanation. The ovule occupies from eight to ten days in its passage from the ovary to the uterus, and it may be impregnated at any time within that period, provided it meet with fertilizing material. If, as maintained by Williams and others, the ovule is discharged at the commencement of a menstrual period, rather than at or near its termination, we can still understand how a coitus taking place within a few days prior to the flow might be fruitful as well as one had within eight or ten days subsequent to a period. But when a single intercourse takes place from twelve to twenty-two days after a menstrual period, or ten days before the next, and becomes fruitful, we cannot accept the explanation given by the ovulationists without additional and different facts.

"To meet the obvious difficulty here presented, it is urged that the spermatozoa may live a long time—indefinitely indeed —in the generative passages of the woman. While we are not able to say positively that such is not the truth, we do say, that so far as we have actual knowledge on the subject, the tenor of life in the spermatozoa is quite limited. Dr. Sims,* who made many examinations of the semen in order to determine how long the spermatozoa may retain their vitality in the matrix, found none alive at a longer period than forty hours, although he admits that they may live longer under favorable circumstances. And he quotes Dr. S. G. Percy as reporting a case in which he found 'living spermatozoa, and many dead ones' issuing from the os uteri eight and a half days after the last sexual connection. If we admit the correctness of all these statements we have no right to assume the persistence of vitality in the

---

* Uterine Surgery, p. 374.

human spermatozoon for a longer period than that given. Granting this term of vitality,—which I feel assured must be quite exceptional,—let us see whether it is sufficient to meet the requirements of some of these cases of impregnation following a single coitus. For example, Montgomery* reports a case in which the last menstruation occurred on the 8th of October. Insemination took place on the 10th of November; pregnancy resulted. Now, if the ovule impregnated were shed at the last menstrual period, twenty-three days must have elapsed between that time and insemination. We cannot suppose the ovule to have retained its vitality and capability of impregnation during this long period, for such a supposition is quite at variance with all observed facts both as regards it† and the history of the decidua (Aveling, Williams, Engelmann). On the other hand, if we suppose that the semen remained in the generative tract until an ovule belonging to the next period was extruded, we must suppose the spermatozoa to resist the mechanical washing away by means of the menstrual flow—a highly improbable notion, and one not made more reasonable by the fanciful idea that the uterus by a sort of instinct, anticipates what is going to take place and governs itself accordingly.‡ Likewise, it implies the vitality of the spermatozoa for a period of eight days, *plus* the time neccessary to meet the descending ovule—probably four or five days more.

"It is well known that many women continue to menstruate, with entire regularity, for a considerable time prior to the final cessation without conceiving; and I believe that this fact is explainable by the gradual failure of the ovaries to furnish perfectly developed ovules. Indeed, it is quite probable that all ova which are thrown off are not capable of impregnation

---

* Signs, etc., of Pregnancy, p. 258.

† "How long after its maturation the ovum can retain its vitality and susceptibility to the seminal influence is not known, but probably the time is short."—DUNCAN, Fecundity, Fertility and Sterility, p. 428.

‡ "Under such circumstances menstruation often does not take place at all, or only very scantily; the uterine system, as it were, anticipating the conception and preventing the failure which might result from a free discharge of blood."—DUNCAN, Fecundity, Fertility and Sterility, p. 431.

at any period of life; for, where other conditions are apparently equal, some females are impregnated every twelve or thirteen months, others every eighteen months or two years, while others have still longer intervals of rest. Dewees mentions an instance of a lady who conceived every seven years, and who bore four children at that interval; and I know one who had a lapse of three years between each of six successive pregnancies. It would seem, therefore, that it requires a certain period to perfect an ovule, and that the time required is much greater in some instances than in others. And if menstruation is produced by ovulation, it appears scarcely probable that a succession of imperfectly developed ovules—so imperfect, indeed, as not to be susceptible of impregnation, or even of extrusion, as we have seen is frequently the fact—should yet be sufficient to maintain a completely normal monthly flow.

"Finally, it is not at all uncommon to find the menses suppressed for some months immediately after marriage, without the occurrence of pregnancy. Are these cases to be explained by supposing that marriage suppresses or retards the development of Graafian follicles?

"From all the foregoing considerations, it seems to me conclusive that ovulation and menstruation may, and frequently do, occur independently of each other; that while they may be coincident, there certainly is no such constant connection between the two as to warrant the assertion that 'at every menstrual period a matured ovule escapes from the ovary'— an assertion which embodies the very essence of the ovular theory.

"2. PHYSIOLOGICALLY, THE PERIOD OF THE MENSTRUATION IN WOMAN CORRESPONDS WITH THE ŒSTRUS OR RUT OF OTHER MAMMALIA.

"It is well known that during certain periods, the intervals between which vary in different species of mammalia below man, ova are matured and extruded from the ovary, and that this process is attended by great excitement of the entire generative apparatus. Upon the supposed similarity of this function—termed rut, œstrus, or œstruation—to menstruation in the human female is based one of the strongest arguments in favor of the ovular theory. Indeed, Cazeaux* and Pouchet† lay especial stress upon it.

"Down to the time of Martin Barry it was believed that sexual congress was the essential determining cause of the rupture of the Graafian follicle, but the experiments of Bischoff, Coste, Pouchet, and others, proved that such rupture was spontaneous and entirely independent of male influence of any kind, both in man and the lower animals, although it was hastened in some instances by coitus. Reasoning, then, from the known analogy existing between this and many other of the vital processes in the lower mammalia and the corresponding ones in man, it was assumed that the conditions of rut and menstruation were analogous, and had the same significance.

"While it is true that many physiological conditions in man and the other animals of the order to which he belongs are subject to the same general laws, these conditions differ in specific points just as much as do the different genera and species of that order in their anatomical features.

"It may be well in this connection, and before enumerating the important points in which œstruation and menstruation differ, to call attention to the fact, that even in those cases in which the œstrus and ovulation are synchronous, it has never been proven that the former is caused by the latter. Indeed, it is far more probable that they are both the result of a common cause—some erethism of the system resulting in congestion and excitement of the entire sexual apparatus.

"The appeal to comparative physiology by the ovulationists has always seemed to me an unfortunate one, for the noteworthy differences between œstruation and menstruation are quite sufficient, I think, to stamp the two processes as wholly dissimilar. These points of difference are as follows:

"1. When, during the œstrus, there is a discharge from the genitals (which is not always the case), it is mucous in char-

---

* Second Amer. ed., p. 9.
† Théorie Positive de l'Ovulation Spontanée, p. 227.

acter, and its source is chiefly the glands of the external organs: its object is to lubricate the parts, and, in some instances, by its odor to attract the male. In woman the discharge is blood, from vascular rupture; its seat, the mucous lining of the body of the uterus, and its presence an indication of the disintegration of that structure.

"2. The excitement characterizing the œstrus is the only period during which the male is received, and the only time when impregnation is possible. In woman, while pregnancy is possible at any time, it usually occurs during the period of rest, that is, in the intermenstrual period.

"3. On the subsidence of the œstrus, there is a period of inappetence, during which the female not only no longer invites, but successfully resists the male approach. At the corresponding time in women, sexual desire is commonly increased, and in some, present at no other time.

"4. The œstrus, or period of sexual desire, is necessary in the lower mammalia, for the reproduction of the species. In women, desire is not essential either for intercourse or impregnation.

"5. Œstruation and ovulation in many animals are determined by changes in the seasons and other surrounding circumstances,* and in some animals (deer) the semen is only elaborated at such times. In man, changes of season, etc., produce no such effect, and semen is secreted constantly.

"6. The œstrus may be excited in some animals (the mare) by the importunities or teasing of the male. Menstruation is neither excited nor hastened by the presence of the male; on the contrary, undue excitement of the generative organs, or of the sexual passion, seems frequently to have a tendency to arrest it, as witnessed in newly married women.

"7. During the œstrus, both the male and female evince a desire for copulation. During menstruation, the female has a delicate shrinking from the act, and the male likewise feels

---

* Barnes, Diseases of Women, p. 148, states that in the wild state the rabbit has only one or two litters a year, but when its young are taken away at a suitable time, it has perhaps seven. So likewise the period of ovular maturation is changed in the case of the pigeon, domestic hen, etc.

more indifference than at any other time, amounting in many cases to positive repugnance.

"8. The ovaries in the lower animals contain ripe ova *only* at the period of heat (Bischoff). In the human female, ripe ova are found at all times without reference to the period of menstruation.

"The foregoing points of dissimilarity are so distinctive, and refer to such important features, that I feel warranted in denying that œstruation and menstruation are corresponding processes.

"3. THE REMOVAL OF THE OVARIES IS AT ONCE FOLLOWED BY CESSATION OF MENSTRUATION.

"Percival Pott, Cazeaux, Wells, Battey, and others, have reported cases in which the artificial removal of the ovaries was followed by the immediate and permanent cessation of the menstrual function; and these facts have been cited to prove the necessity of these organs for the maintenance of the periodical flow.

"It must be admitted that, if such an effect were the constant and certain result of double ovariotomy, it would go far towards showing the necessity for ovarian influence. But such result is not constant; indeed, the instances in which both ovaries have been removed without interruption or discontinuance of the menstrual flow are so numerous and authentic, that recent writers, who, like Leishman, affirm as an admitted fact 'the invariable and immediate cessation of menstruation when the ovaries have been removed,' subject themselves fairly to charges either of ignorance or want of candor.

"Dr. John Goodman* has compiled the following table showing the effect of double ovariotomy upon the menstrual function in all the cases of which he could obtain information down to the year 1872:

---

* Richmond and Louisville Med. Jour., Dec. 1875.

INTRODUCTION. 41

"*Table of Cases in which both Ovaries have been successfully removed from Women under Forty-five years of Age.*

No.	Operator.	Quoted from.	Date.	Age	
1	Pott, ...	A. J. Med. Sci., 1844,	178–	23	
2	J. L. Atlee, . .	A. J. Med. Sci., 1844,	1843	29	
3	Bird, ...	Lancet, 1848, ...	1847	32	Menstruation uninterrupted; tendency to menorrhagia.
4	Peaslee, . . .	Lyman's Table, . . .	1850	24	
5	Burnham, . . .	Lyman's Table, . . .	1853	42	
6	W. L. Atlee, . .	Atlee on Ov. Tumors, .	1854	35	Menstruation regular. Ceased in 1864, forty-fifth year.
7	W. L. Atlee, . .	Atlee on Ov. Tumors, .	1855	19	Regular molimen, with white discharge.
8	W. L. Atlee, . .	Atlee on Ov. Tumors, .	1861	41	Menstruation regular to 1863, when last reported.
9	Peaslee, . . .	A. J. Med. Sci., 1863, .	1862	35	
10	Peaslee, . . .	A. J. Med. Sci., 1864, .	1863	39	
11	W. L. Atlee, . .	Atlee on Ov. Tumors, .	1864	34	Last report 1870. Menstruation regular to that time.
12	Beattey, . . .	Wells, Dis. of Ovaries,	1865	37	
13	Storer, ...	A. J. Med. Sci., 1864,	1866	. .	Menstruating regularly a year after operation.
14	Storer, ...	Peaslee on Ov. Tumors,	1867	43	
15	Wells, ...	Wells, Dis. of Ovaries,	1868	39	
16	Wells, ...	Wells, Dis. of Ovaries,	1869	22	
17	Hicks.	Wells, Dis. of Ovaries,	1869	39	
18	Monro, ...	Wells, Dis. of Ovaries,	1870	34	
19	Mayer, ...	Wells, Dis. of Ovaries,	1871	29	Last report one year after operation. Menstruation regular.
20	Meadows, . . .	Lancet, 1872, ...	1871	. .	Last report 6 months after operation. Menstruation regular.
21	Priestly, . . .	Wells, Dis. of Ovaries,	1872	22	Continued to menstruate to the forty-seventh year of her age.
22	A. R. Jackson, .	Peaslee, Ov. Tumors, .	1865	44	Menstruation regular.
23	Le Fort, . . .	Peaslee, Ov. Tumors, .	. .	. .	Menstruation, but not regularly.
24	Baker Brown, .	Peaslee, Ov. Tumors, .	. .	. .	Menstruates regularly from cicatrix and vagina.*
25	Baker Brown, .	Peaslee, Ov. Tumors, .	. .	. .	Menstruation regular.
26	Koeberle, . . .	Peaslee, Ov. Tumors, .	. .	. .	Menstruation regular.
27	Battey, ...	Personal information, .	1872	23	Irregular sanguineous discharges, sometimes profuse.

Clay, of Manchester, had four cases in which there was subsequent sanguineous discharge — (Peaslee).

"Dr. Goodman says: 'In order to determine as accurately as possible the effects of the removal of both ovaries upon the menstrual function, I have carefully examined and arranged all the cases of which I could obtain reports; . . . . irregular sanguineous discharges, I have, of course, not counted as menstrual.

"'Of the twenty-seven cases here recorded, it will be observed that in nearly one-half menstruation was not affected by the removal of the ovaries; in one the hæmorrhagic discharge

---

* The whole uterus, except cervix, removed with ovaries.

was increased; in one it was diminished; and in several sanguineous flows occurred at irregular intervals.'

"Dr. Ely McClelland, of Louisville, Ky., in a private letter, dated March 18th, 1876, gives the following facts bearing upon this point, and kindly places them at my disposal. The cases referred to were operated upon for pernicious ovulation, by 'Battey's operation,' or that known as 'normal ovariotomy.'

"'CASE I.—But one ovary was removed.

"'CASE II.—Both ovaries were removed, one in May, 1875, and the other in September, 1875. This lady has regularly and persistently menstruated since the operation.

"'CASE III.—Both ovaries removed in August, 1875. This case had menstruated vicariously prior to the operation, and is still the subject of such disorder.

"'CASE IV.—The ovaries of this lady were removed in September, 1875. She menstruates regularly.

"'So far as these Louisville cases go, the removal of both ovaries, after the menstrual function has been established, produces no influence upon the regularity of its occurrence. What may result after the lapse of a few more months, it is of course impossible to determine.'

"Indeed, it is so well known that the removal of the ovaries does not necessarily induce the menopause, that many of those who formerly denied the fact now admit it; but they endeavor to explain the circumstance consistently with the ovular theory. Some of these allege that the ovaries are not the only source of Graafian vesicles. Spencer Wells, for example, states* that occasionally the essential elements of the ovaries are sometimes scattered between the layers of the peritoneum, as in the lower animals; and that in some cases 'Graafian follicles have been seen developing in some of the mammalia at a distance from the entire ovary, and that such vesicles have developed into unilocular tumors.' Sappey† likewise states that it is not rare to find a score or more cystic ovules, some of them the size of a pea, on the alar mesentery, in the neighborhood of the ovary, and he accounts for their presence in this

---

* Diseases of the Ovaries, p. 11.
† Quoted by Savage, on the "Female Pelvic Organs."

unnatural situation by supposing that they 'failed to reach their destination owing to some abnormal relations on the part of the Fallopian tube.'

"Now, while it would be presumptuous to deny that such a condition of things as that mentioned by the last-named author is possible, surely it must be exceedingly rare; and, so far as I am aware, there is no instance in which such an anomaly has been found in the human female. I apprehend, therefore, that not the most ardent advocate of the ovular theory would be willing to advance such a hypothetical circumstance to account for the appearance of a periodic monthly hæmorrhage in thirteen of twenty-seven cases of removal of the ovaries. For such an argument would involve the absurd assumption that an ovum which had failed to reach the uterus after maturation and extrusion, could return to the immature condition and ripen over again, and that, too, without its enveloping fluid and capsule!

"The condition mentioned by Mr. Wells must be equally rare, and seems equally weak as a foundation for an argument. In regard to both of these conditions, Dr. Goodman, in the paper already referred to, says: 'I think it a very fair conclusion that if such vesicles really existed, they were totally extirpated in some, if not the greater part, of the thirteen cases in which menstruation continued after the removal of both ovaries. Even if some of them remained, it is clearly impossible that they could have been sufficiently numerous to have afforded a ripened vesicle every month for ten or more years. Their only effect would have been to stimulate the nervous system, and maintain in a more perfect degree the ovarian development.'

"Others again have explained the persistence of menstruation after extirpation of the ovaries by force of habit. Schroeder* expresses the argument thus: 'We prefer in such exceptional cases, instead of drawing the conclusion which is directly opposed to all our views, viz., that menstruation has absolutely nothing to do with the presence of ovaries, to assume

---

* Diseases of the Female Sexual Organs, p. 318.

that in these women, too, menstruation was caused by the growth of Graafian follicles in their ovaries, but that the organism had, in the course of years, become so accustomed to the regular discharge of blood that this still continued, although the ovaries were removed.'

"But, surely, this is no explanation; it is nothing more than a reiteration of the fact in other terms. The 'habits' of our bodies are not causeless: they are all explainable on a rational basis. No act is performed in the animal economy without some antecedent cause, and the same may be said of every recurrence of such act. In the case of the menstrual flow, if its periodicity were maintained in the past by the successive evolution of ovules, such ovular action would be necessary still; and if the cause ceased at any time to act, so likewise would the effect cease. But even this alleged force of habit fails to meet the facts in a case reported by the writer in the *Chicago Medical Journal*, for October, 1870, and which appears in Dr. Goodman's table as No. 22. In this case, the patient. forty-four years of age, had both ovaries, together with a portion of Fallopian tubes, removed. A menstrual period had ceased on the 30th August, 1865: the operation was performed the following day. On October 1st, thirty-one days afterwards, a sanguineous discharge appeared and lasted four days. attended by the usual symptoms of menstruation—lassitude, nervousness, backache, etc. There now occurred an interval of eighty-three days, the discharge reappearing December 22d. Its next appearance was on January 20th, 1866,—four weeks after,—and from this last date it continued to return with entire regularity every twenty-eight or twenty-nine days, attended by all the ordinary menstrual accompaniments, and lasting each time from three to five days, down to October, 1867—a period of twenty-two months. It then ceased until February. 1868, when it appeared for the last time, the lady being then forty-seven years of age.

"In this instance, the interval of nearly three months, during which the discharge was absent, was certainly sufficiently long to break up any mere habit, and shows that we must look to some other impelling force in order to account for the subsequent return to regularity. Here were no ovaries, no monthly developing ovules, an interruption for nearly three periods of the menstrual 'habit,' and yet menstruation returned and continued regularly to reappear down to the normal time of final cessation.

"The facts of periodicity in the human body are more numerous than generally supposed, and most interesting in their character. Without any intention of amplifying upon the subject, I will merely remark that it is now universally admitted that all forms of periodicity, whether of a physiological or pathological character, depend upon the nervous system; and there are numerous facts which warrant us in narrowing this dependence still further, and limiting it to a particular division of the nervous system—the sympathetic. It is well known that a frog's heart will continue its regular systole and diastole a considerable time after its removal from the thorax. The only motive agency left to it then, so far as we know, is that furnished by the sympathetic ganglia which are imbedded in its substance. The influence of these centres of nervous action is neither continuous nor occasional, but rhythmical—that is, periodic. The uterus resembles the heart in also possessing numerous sympathetic ganglia imbedded in its walls, and in being wholly independent of the cerebro-spinal system in its movements. Furthermore, the recent researches of Goltz and Freusberg* seem to show that there exists in the lumbar portion of the spine a nervous centre for the sexual functions. These facts and investigations may afford a clue to the explanation of the persistence of sexual appetite and functional activity of the generative organs after the destruction by disease, or removal of the ovaries—although in these organs undoubtedly originates the primary impelling force which sets this complex sexual machinery in motion. But whatever may be the nature or exact seat of this force, I believe that its action must be persistent and, in a sense, continuous, although some of its results be rhythmical. A pendulum may

---

* See Dr. Duncan's Address, Obstet. Journal, Great Britain and Ireland, 1875, p. 361.

be set in motion by a single forcible impulse, and for a time it will continue to swing; but unless the application of the force be continuous or repeated, the arc described by the moving body will become shorter and shorter, until finally the motion will cease wholly.

"According to the light thrown upon the subject of menstruation by the latest researches, we are perhaps justified in propounding the following as embodying the main facts:

"The reproductive organs of the female, including the ovaries, Fallopian tubes, uterus, and vagina, receive their vascular and nervous supplies from the same sources. Prior to the age of puberty, all these organs are in a state of comparative quiescence, and the uterus of a girl of eleven or twelve years is scarcely larger than the organ in infancy. Notwithstanding the fact that ova undergo some degree of development and are discharged from the ovaries from early childhood onward, their growth proceeds slowly, and, so to speak, unperceived by the nervous system. At or about the fifteenth year, the uterine mucous membrane attains a high degree of development, and, at the same time, the erectile tissues of the other genital organs, external and internal, arrive at their structural completion. Like a wound-up clock, with its needed touch to the pendulum, these organs now only wait for some sufficient impulse to arouse them to functional activity. This is afforded by the next recurring period of ovulation. By the advancing growth of one or more vesicles, an irritation of the ovarian nerves is produced; the effect of this upon the sympathetic, and, by reflex action, upon the vaso-motor nerves, is an increased hyperæmia in the uterus and other genital organs. The uterine congestion thus produced especially affects the lining membrane of the organ, for the reason that, structurally, it is more liable to vascular turgescence than the parenchyma. This vascular activity is followed by a corresponding increase of nutrition and hypergrowth—this latter consisting both in multiplication of the cellular elements of the parts and development of those already existing. The superficial vessels of the membrane are greatly enlarged around the glandular orifices, as are also the glands themselves. The entire membrane is so thickened and convoluted that the uterine cavity seems scarcely large enough to contain it. This process—called 'nidation' by Aveling—takes place in order to supply the possible needs of an impregnated ovum, and should such a one reach the uterine cavity, the developed membrane becomes its future nidus. But if not, a retrograde metamorphosis now takes place. The supergrown parts of the uterus, consisting, as already stated, chiefly of the mucous membrane, lose their excess of blood-supply and die of starvation. The first elements which suffer death are the epithelial cells which line the mucous membrane; next, the new cells of the connective-tissue stratum below; and, finally, the vessels which are developing or may have developed from this surface. All these parts become infiltrated with fat, the new formations are carried off, the vessels open, and there results the active hæmorrhage which constitutes the menstrual flow. This process is repeated at regular intervals corresponding to the periodic life of the individual, and varying somewhat in different cases.

"In the sense and to the extent just indicated, I regard ovulation as necessary to menstruation; it furnishes to the structurally completed uterus, through the medium of the ganglionic nervous system, a needed hyperæmia to *originate* the menstrual discharge. In order to do this, it is not necessary that a follicle should burst (Ritchie), although it may do so. Indeed, I have no doubt that a follicle may pass through several periods without discharging its contained ovule. Doubtless, the pelvic congestion of the menstrual period greatly stimulates the maturation of the follicle, just as does the excitement of sexual intercourse, and to a much greater extent, probably, because of its longer continuance; and a follicle which has been subject to these successive periods of excitement eventually matures and bursts, with, perhaps, an occasional exception. The increase in its fluid contents, the thinning of its walls, and its near approach to the surface of the ovary, all conduce to its easy rupture and such rupture may occur at any time, although it is clearly more likely to do so during the menstrual congestion, the excitement of intercourse, or when on the point of bursting, from a blow on the abdomen (Schroeder). Such is probably the usual course, although, as already intimated, not an invariable one; for all authorities admit that some follicles never attain full development, but after arriving at a certain stage of growth, cease to enlarge and finally shrink and disappear.

"Menstruation, with its phenomena of regularly recurring development and disintegration of the uterine mucous membrane, once established, proceeds side by side with the process of ovulation. The two, while accompanying and aiding each the other, are yet mutually independent; and menstruation, instead of being an effect of ovular maturation and dehiscence, is rather, in a certain sense, their cause. In menstruation, the organ chiefly, and the only one essentially employed, is the mucous membrane of the body of the uterus; the other pelvic organs, that is to say, the uterus proper, the ovaries, Fallopian tubes and vagina, have no part in the process beyond their share in the attendant general pelvic congestion (Beigel).

"There are many facts connected with menstruation which are not satisfactorily accounted for by the ovular theory, and I desire, in conclusion, to call attention briefly to a few of them.

"(a.) The first is that variety of the function known as *remittent*, where the habitual type is changed to another in which the flow occurs usually at shorter intervals—every fortnight, for example. These cases are strongly antagonistic to the received theory. Tilt* considers them as dependent 'upon some perversion of the nervous force presiding over the generative function, because those in whom the anomaly is observed are generally of a delicate and nervous temperament,' and also because he has always succeeded in restoring menstruation to the monthly type by the exhibition of quinine, a remedy whose efficacy in controlling nervous derangement

---

* Uterine and Ovarian Inflammation, third Lond. ed., p. 205.

of a periodical character is well known. Negrier,* quoted by Tilt, after observing that several patients did not suffer from fortnightly menstruation, says: 'I do not believe that the ripening of the ovarian vesicles can take place in less than a month; so, in these cases, I think it more natural to suppose that the two ovaries might so progress monthly that, for instance, the right would contain a ripe vesicle on the first of the month, while in the left ovary a vesicle would ripen on the fifteenth.' Does any one regard ovulation as a process whose type of periodicity would be changed by the administration of quinine?

"(b.) Again, the ovular theory does not account for the regular recurrence of menstruation after the removal of one ovary. Ovariotomists are unanimous in the statement that in cases where a single ovary is removed, a healthy one being left, menstruation is not interrupted, or, at least, the function is no more deranged than it would be by any other equally severe surgical operation.

"It is not known how the work of maturing ovules is divided between the two ovaries. Mr. Girdwood, from observations made in several cases, states that the number of cicatrices found in the ovaries corresponded with the known number of menstrual periods, and that they were equally distributed between the two organs. Others think that for many months in succession one ovary may furnish all the ovules, and then remaining quiescent, that the other assumes the work for an equal, or possibly an unequal length of time. But it is plain that if either of these hypotheses be accepted—if ovulation be regular in any manner, and its periodicity depend upon the presence of both ovaries, it would be interrupted necessarily by the removal of either of these organs. In the first case it would occur only at intervals of two months; and, in the second, according as the active or quiescent organ, for the time being, were removed, we should have an entire temporary cessation at once, or at the end of

---

* This author believed that the ovaries alternated their action, one furnishing an ovule one month, and the other the next. This is likewise the opinion of Girdwood.

its term of activity; and if menstruation were dependent upon ovulation, a corresponding aberration of regularity would be observed in it also. But this never takes place, for, as already stated, single ovariotomy is, as a rule, followed by no change whatever in the menstrual periodicity.

"It cannot be said, in answer to this, that, as in the case of the kidneys or testicles, the removal of one gland is followed by increased and compensating work by its fellow. The ovary is not a gland, and the Graafian vesicle is not a secretion. The office of the ovary is simply and only to furnish a suitable place for the development of the primordial follicles existent in its stroma from the beginning.

"(c.) The remarkable regularity in the ripening and discharge of ovules, one after another, month after month, which is assumed by the ovular theory, is combated also by the frequent occurrence of a simultaneous discharge of two or more ova.* Multiparous pregnancies can, I think, only be rationally accounted for by the fact that the shedding of ovules is an irregular function, proceeding in both ovaries simultaneously and independently.

"(d.) The ovular theory wholly fails, too, to account for the menstrual irregularities caused by mental influence. Tilt† says: 'I have patients in whom any unusual nervous emotion or overexertion will bring on the menstrual flow, with the usual menstrual symptoms, although they may have only just recovered from this discharge. How can it be supposed that an ovule can be ripened, and the dense ovarian envelope suddenly perforated, by the fatigue of a dinner-party, by hearing disagreeable news, or by an altercation with a servant?' It is well known that influences such as those just mentioned may cause the discharge to appear, and may equally check it when present; and it is likewise known that these, or other similar disturbing causes, may at once change the menstrual regularity; the flow appearing after the usual interval, whether the last one occurred at the right or wrong

---

* Ritchie, in a case reported by Cazeaux, found six ova in the uterus at one time.

† Change of Life, p. 72.

time. 'This sudden shifting of periodic action is the special attribute of the nervous system; it shows the menstrual flow to be impelled by nervous influence, and explains how a strong emotion may repel it or alter the time of its appearance.' (Tilt.)

"The argument which I have endeavored to make may be thus summarized:

"1. Ovulation and menstruation may each occur independently of the other.

"2. Ovulation is the irregular but constant function of the ovaries, while menstruation is the regular rhythmical function of the uterus (Kesteven).

"3. Ova are matured and discharged from the ovaries at all periods of female life, from early childhood to old age, both before puberty and after the menopause; hence the one cannot be the sign of the other.

"4. Menstruation is the consequence of conditions established by the structurally completed uterus, and depends upon ovulation *only* for its origination.

"5. The mucous membrane of the uterine body is the only organ essentially concerned in the menstrual act; the uterus proper, the ovaries, Fallopian tubes, and vagina have their functional activity increased, however, by receiving a share of the general pelvic congestion which accompanies the process.

"6. The menstrual congestion of the pelvic organs—of the ovaries in particular—is, of all causes, the one most likely to determine the ovipont when a Graafian vesicle is sufficiently mature, and hence ovulation and menstruation are frequently concurrent.

"7. The theory that would make menstruation dependent upon ovulation fails to account for the possible occurrence of pregnancy at any and all times between the menstrual periods; for multiparous conceptions; for the frequent persistence of menstruation after the removal of *both* ovaries; for the noninterference with menstrual regularity by removal of *one* ovary, and for the menstrual derangements and the shifting of menstrual periodicity from mental emotion.

"8. All the known facts in regard to both ovulation and menstruation are consistent with the theory that, after the latter is once established, the two functions proceed side by side, but independently of each other, the former occurring at irregular and the latter at regular intervals; while, on the contrary, many of these facts are wholly inconsistent with the theory that assumes a necessary ovular maturity and rupture at each menstrual period."

CAUSES OF STERILITY.

CHAPTER I.

I WILL now proceed to enumerate the causes of sterility, viewed in the light of the present physiological and anatomical knowledge of the generative organs of woman.

With a full appreciation of the imperfections which it contains, I submit the following arrangement or classification.

I. CONSTITUTIONAL OR PREDISPONENT.
    a. Obesity.
    b. Chlorosis.
    c. Scrofula.
    d. Syphilis.
    e. Mercurialization.
    f. Twin birth.
    g. Prostitution.
    h. Inordinate sexual intercourse.
    i. Change of climate.
    j. Mineral waters.
    k. Improper diet, clothing, and exercise.

II. PSYCHICAL.
    a. Incompatibility.
    b. Frigidity.
    c. Erotism.

III. OVARIAN.
    a. Atrophy of the ovaries.
    b. Absence of the ovaries.
    c. Imperfect development of the ovaries.
    d. Inflammation—chronic—of the ovaries.

c. Degeneration of the ovaries.
f. Tumors of the ovaries.
g. Dropsy of the ovaries.
h. Dislocation of the ovaries.

IV. UTERINE.
1. Fallopian tubes.
    a. Stricture.
    b. Obstruction.
    c. Inflammation.
    d. Inflammation of fimbriated extremity.
    e. Displacement of fimbriated extremity.
2. The Uterus.
    a. Absence of.
    b. Atresia; constrictions; occlusions of the uterus.
        Atresia; constrictions of cavity and cervix.
        Atresia; constrictions; acquired, congenital.
        Imperfect development of uterus.
    c. Atrophy.
    d. Displacements.
        Anteflexion, and anteversion.
        Retroflexion, and retroversion.
        Lateroflexion.
        Prolapsus.
        Elevation.
        Inversion.
    e. Tumors of.
    f. Inflammations, chronic.
        Endometritis.
        Endocervicitis.
    g. Ulceration.
    h. Leucorrhœa.
    i. Amenorrhœa.
    j. Dysmenorrhœa.
    k. Menstrual irregularities.
    l. Abortions.
    m. Areolar hyperplasia.
    n. Abnormal shape of the cervix.

V. VAGINAL.
  a. Atresia.
      Congenital.
      Accidental.
  b. Absence of vagina.
  c. Non-retaining vagina.
  d. Imperforate hymen.
  e. Vaginismus.
  f. Leucorrhœa.

VI. RECTAL.
  a. Hæmorrhoids.
  b. Prolapsus ani.
  c. Fissures of the anus.

VII. MEDICINAL.*

Agnus castus.	*Conium.*	*Mercurius.*	*Ruta graveolens.*
Apis mel.	*Cantharides.*	*Morphine.*	*Sumbul.*
Asarum.	*Chimaphila.*	*Moschus.*	*Senecio.*
Baryta carbonica.	*Ferrum.*	*Sepia.*	· *Sabina.*
Cimicifuga.	*Iodine.*	*Phytolacca.*	*Secale cornutum.*
Capsicum.	*Iodide of Lead.*	*Phosphorus.*	*Trillium.*
Caladium seg.	*Kali brom.*	*Platinum.*	*Ustilago.*
Caulophyllum.	*Lachesis.*	*Plumbum.*	
Cannabis.	*Iodide of barium.*	*Pulsatilla.*	

* The above enumeration of the medicinal causes is necessarily imperfect, because much of the testimony relating to their sterility-causing power is unreliable. Their claims will be separately considered in another place.

CHAPTER II.
STATISTICS OF STERILITY.

In general, it is stated by Sims and other authorities, that *about every eighth marriage is sterile.*

This is certainly an astonishing statement. In no other age of the world than this could such a state of things exist. From what we can learn from ancient writers, both sacred and profane, sterility was a very uncommon occurrence among the Egyptians, Jews, Greeks, Romans, and the nations of Europe. If all the other evidence was lacking, the fact that sterility was considered a disgrace among the nations enumerated above would be sufficient to prove its extreme rarity among their women.

In Dr. Duncan's work on *Fecundity, Fertility and Sterility,* we find that in the cities of Edinburgh and Glasgow, 15 per cent. of all the marriages between fifteen and forty-four years of age are sterile. Dr. Lever (*Org. Dis. of Uterus,* p. 5) says: "It is found that $\frac{1}{20}$, or 5 per cent., of married women are wholly unprolific." Duncan does not ascribe much value to this estimate. Dr. Simpson (*Obstetric Works,* vol. i, p. 328) says it has been observed that in some portions of Sweden one barren woman is not met with for ten fertile.

Frank asserts that only one marriage in fifty is unproductive, but he gives no proof of his estimate.

Thompson (*Todd's Cyclopedia,* vol. ii, p. 478) estimates from a census of two large cities, Grangemouth and Bathgate, that 1 in every $10\frac{1}{2}$ married women were sterile. The same author gives the analysis of 503 marriages in the *British Peerage,* resulting as follows: Among 495 marriages 81 were unproductive, or 1 in $6\frac{1}{5}$ were without any family.

Dr. West (*Diseases of Women,* p. 3) states that he found the general average of sterile marriages among his patients at St. Bartholomew's Hospital to be 1 sterile marriage in every 8.5.

Dr. Duncan's tables show that about 7 per cent. of all the marriages between 15 and 19 years of age, inclusive, are without offspring. Those that are married at ages from 20 to 24, inclusive, are almost all fertile; and that, after that age, sterility gradually increases, according to the greater age at the time of marriage.

We have no means of ascertaining the exact proportion of sterile women among the older nations. Even at this day statisticians have been remiss in collecting, or unable to obtain any reliable tables. Dr. Nathan Allen* reports to the Social Science Congress that he has ascertained that in New England the birth-rate of the strictly American class is steadily diminishing.

From a careful examination of the census and registration reports, as well as of town and city records, the following facts were developed:

The average number of births to each married couple, and the relative number of children to adults, with the present generation of Americans, did not begin to compare with what it was in New England one or two hundred years ago.

A census of Massachusetts, taken in 1765, reports almost one-half of the population as under sixteen years of age; whereas, from the best estimates that can be made, not more than one-fourth, or, perhaps, one fifth, of the present American population are under that age. . . . . An examination of the records also shows that the average number of children to each family, at the date above mentioned, was from eight to ten; and that from the second, third, and fourth generations the average ranged from seven to eight; but when we come to the fifth and sixth generations the number of children diminished more rapidly, averaging only four or five to each married couple; and that of the present generation, the number will not much, if any, exceed three children.

Now, if we take into account the known increase in miscarriages from disease, as well as the greater frequency of criminal abortions, we shall find that this decrease in the

---

* Psychological Journal, vol. ii, p. 214.

birth-rate is due largely to *the increase in the number of sterile women.*

It must be borne in mind that many of the causes of miscarriage are also causes of sterility. A diseased condition causing the former in one woman, will in another cause the latter. The statistics of sterility are, therefore, closely connected with those of abortion.

Now it is generally believed that miscarriage from causes not designated as criminal is one hundred times more frequent now than a few centuries ago; so we may safely assert the same of sterility.

If, therefore, the present rate *is one in eight,* the ancient rate would be nearer *one in eighty.*

The statistics of sterility, as connected with certain diseased conditions, have been only partially arrived at. We can, therefore, only give a few estimates.

As regards the statistics of sterility connected with *uterine displacements,* Sims gives the following table, which shows the influence displacements of the uterus exercise on the sterile condition:

Of 250 cases of *natural* sterility, 103 had anteversion; 68 had retroversion.

Of 255 cases of *acquired* sterility, 61 had anteversion and 111 had retroversion.

Or out of 505 cases of sterility, 343 had uterine displacements, or about two-thirds of the whole.

The statistics of sterility connected with *fibroid tumors of the uterus* show that of 225 women who had once borne children and then become sterile, 38 had fibroid tumors of various sizes and variously seated, or 1 in 6.7.

Of 250 women who had *never* borne children, the cause of sterility was found to be complicated with the presence of fibroid tumors in 57, being at the rate of about 1 in 4.3.

In 100 virgins consulting for some uterine disease, 24 had fibroid tumors, or 1 in 4.6. These latter would probably be sterile.

Hewitt (*Diseases of Women,* p. 238), gives the following: "*Influence of acquired deformities of the uterus in producing Sterility.*"

Of the 296 cases 235 were married or had had children, including 100 cases of retroflexion and 135 cases of anteflexion.

"Of the 235 cases 81 were sterile, in the sense that they had either had no children or had only had abortions.

"Of these 81 cases 57 were absolutely sterile, and 24 had only had abortions.

"In a very considerable number of cases the patients had had one or more children, but had been subsequently sterile. These cases are not included in those just described, but their importance is equally great, and the interference with *subsequent* procreation of children was evidently connected with the presence of a deformity of the uterus acquired subsequently to the first pregnancies. An additional proof, if such were needed, of the effect of the flexion in producing such 'secondary sterility' resides in the fact, presenting itself on looking over these cases, that the symptoms leading the patient to seek for medical advice had very generally existed since the birth of the last child."

CHAPTER III.

CAUSES AND TREATMENT.

I. Constitutional or Predisponent.

CHLOROSIS.—In actual chlorotic conditions conception rarely occurs. If, with the chlorosis, amenorrhœa is present, fecundation is still more difficult.

Treatment.—*Ferrum phos.* is the preparation most valuable in this condition. *Phosphoric acid* is sometimes more useful than any iron preparation, especially when the chlorosis is due to mental causes. *China*, if from loss of fluids, or from latent malaria. *Calcis hypophosphis* is an excellent remedy in many cases. *Nux, Ignatia,* and *Strychnia* are remedies which powerfully stimulate the trophic nerves and thus affect the processes of sanguification. *Pulsatilla, Helonias, Cuprum, Manganese* and *Aletris* are all useful in their sphere.

SCROFULA.—Scrofulous women are oftener fruitful than otherwise. There are cases, however, where the diathesis is well marked, and the ovaries or uterus become the seat of its local manifestations with resulting sterility.

Treatment.—A careful study of Hahnemann's *Chronic Diseases* is essential to the proper selection of remedies for scrofulosis. Among the most important are *Ars., Calc., Hepar sul., Iodine, Cistus, Graph., Kali hyd., Ferr. iod., Merc. iod., Oleum jecoris, Phytolacca, Stillingia* and *Sulphur.*

SYPHILIS.—This malady is doubtless a frequent cause of sterility. Whitehead* narrates a case of "sterility of fifteen years," caused by secondary syphilis.

Treatment.—Syphilis in all its stages and manifestations is

---

* Abortion and Sterility, p. 354.

readily controlled by a few specific remedies, namely, *Mercury com.*, *Merc. iod.*, *Kali hyd.*, *Phytolacca*, *Stillingia*. For many years I have rarely used any other remedy than the *Iodohydrargyrate of potassa*, mentioned in the *New Remedies*, 4th edition, p. 695, vol. ii.

MERCURIALIZATION.—There can be no doubt in the minds of physicians who have closely studied cases of sterility, that Mercury sometimes causes a difficulty of conception.

Treatment.—Mercurial poisoning is successfully treated and the poison eliminated by the judicious use of *Iodide of potassa, Hepar sulph., Kali chlor., Aurum, Nitric acid, Phytolacca, Stillingia*, and *Sulphur*.

OBESITY.—This condition is considered, both popularly and by the profession to be alike the cause and consequence of sterility.

Some authorities doubt if obesity is opposed to fecundation. Why this condition should interefere with conception has never been fully explained. The fact, however, remains, and every physician in large practice has doubtless observed, that sterile women are oftener obese than lean.

When a woman becomes normally sterile, *i. e.*, after the climacteric period, she is quite apt to become adipose. This fact would seem to prove that there is some connection between barrenness and adiposis. This connection is recognized by all breeders of cattle and horses, who consider "fatness" a decided obstacle to conception.

I have known many instances where a woman has had one or two children, and then suddenly become increased in size and weight, with resulting barrenness. These same women, a few years after, from some illness or other cause, would become lean, and again bear children.

Treatment.—Obesity and plethora cannot be safely treated with drugs. In order to diminish the amount of fat or blood in the system by drugs, great damage may, and generally does, result to the general health. The safest medicine for obesity is a decoction of *Seawrack* or *Fucus*. A rigid meat diet, with active exercise, will generally diminish obesity. True plethora, on the contrary, diminishes when a meat diet is abandoned, and light farinaceous food takes its place. (Consult Banting, "*On Obesity*," and Griffin's "*How to Grow Lean*.")

TWIN BIRTH.—"There is a popular belief among a certain class that when a male and female are born at a twin birth, the female will be henceforward sterile. It is probable that this belief originated with the graziers, for it is stated as a fact that when a heifer and bull twins occur, the heifer will not conceive. Among cattle-growers this heifer is called a 'freemartin,' and so undoubted is this belief, that this free-martin is either killed as a calf, or, if raised, is regularly broken to the yoke, and worked like an ox. Many men of intelligence have made this statment to me, and it may be found in agricultural journals."

"Among cattle such may be the fact, although I very much doubt it; but after a careful examination I am convinced that such is not the fact in respect to the human race as I have known of numerous instances to the contrary, and do not now remember ever hearing one statement in corroboration of this opinion."*

Professor Simpson, of Edinburgh, who has made considerable research upon this subject, acknowledges the "infecundity of free-martin cows to be a very general fact, but by no means a universal one." He has also collected the facts relative to the married history of 123 females born co-twin with males, 112 of whom had offspring, and 11 none, although married several years. In other words, the marriages of the females born under the circumstances, "*were unproductive in the proportion of one to ten.*"†

PROSTITUTION.—Prostitutes are notoriously sterile. Any one who will wade through the horrible and sickening details of the effects of prostitution, as described by Sanger,‡ will be con-

---

* Gardener, on Sterility.
† Simpson's Obstetric Memoirs, vol. i, p. 294.
‡ History of Prostitution.

vinced that fecundation rarely takes place in utterly abandoned women. To the moralist this seems to be more a special providence than any physical results. If it were not so, it would be terrible to contemplate the vast increase of crime and misery that would result. According to Sanger, prostitutes who reform and marry, even if not affected with syphilis, are often barren, while others become prolific.

INORDINATE SEXUAL INTERCOURSE.—It is stated by some authorities that this habit, equally with prostitution, may be a cause of barrenness. It has been successfully recommended in such cases, that, in order to insure conception, coitus be indulged in only once during the ten days succeeding the menses. In several cases of sterility I have found this recommendation to be followed by pregnancy. In some cases coitus should be prohibited for a whole month, to allow the exhausted organs time to regain their tone.

CHANGE OF CLIMATE.—According to Casanova, "change of climate will produce sterility, as it has been observed in women born in the region of South America which lies between the rivers Amazon and Nepo. When such women leave their native home to reside in some of the limiting States, they will surely (like many plants, when transplanted from one climate to another) become, if not absolutely barren, at any rate less prolific than in their own native land."

This may be so, but the fact is equally apparent that change of climate often cures sterility. I am informed by the physicians of Cleveland, Buffalo, and New York that women, sterile for years, have conceived while on a visit to the Upper Great Lakes, or a few months' residence at Mackinaw. I have known the same result to follow in cases of ladies from Chicago, who supposed themselves sterile until a visit to Mackinaw or a trip to Lake Superior resulted in a pregnancy. A visit to Europe or the West Indies is often recommended to sterile women, and with good results.

Women sterile in New England become prolific in the West, and women who have never conceived in any part of the country east of the Rocky Mountains become pregnant soon after reaching California.

Baudelocque mentions the case of a high personage who could not have any children by his wife. During an absence of two years from France his wife became pregnant, and gave birth to a boy. After this his wife remained sterile for four years. After another absence from France, she had another child. The change of climate was evidently the cause of some favorable change in the condition of the reproductive organs. This same couple, by making annual visits, became the parents of eleven children, five boys and six girls.

MINERAL WATERS.—"Long residence near chalybeate springs, and the abuse of drinking or bathing in their waters, have been observed, on the continent, to produce sterility in prolific women." (Casanova.) This author believes this result to be due to the *iron* contained in the water, and floating in the air. He accounts for the cure of sterility by *Ferrum* on homœopathic principles. Sterile women are often sent to mineral springs for the cure of barrenness, and it is asserted that some springs are noted for their good effects on this condition.

Among the springs most noted for the cure of sterility are the "Sweet Springs," in Monroe County, West Virginia, and several in Europe.

IMPROPER DIET AND WANT OF EXERCISE.—It is well known to all practical men that the diet has much to do with fecundity in the human subject, also in animals and plants. One of the best writers on this subject—Mr. Doubleday—says: "It is a fact admitted by all gardeners, as well as botanists, that if a tree, plant, or flower be placed in a mould either naturally or artificially made too rich for it, a plethoric state is produced, and fruitfulness ceases. In trees, the effect of strong manures and over-rich soils is, that they run to superfluous wood, blossom irregularly, and almost or entirely cease to bear fruit. In flowering shrubs or flowers the first effect is that the flower becomes double, and loses the power of producing seed; next, it ceases almost even to flower. On the other hand, when a gardener wishes to save seed, he does not give the plant an extra dose of manure, but he subjects it to some hardship, and selects the fruit that is the least fine-looking, knowing that it will be filled with seed; while the finest fruit will be nearly destitute.

"In the animal kingdom, fecundity is totally checked by the plethoric state, while it is induced and increased by the deplethoric or lean state. Rabbits, swine, sheep, and horses, when overfed, will not reproduce; put them out of condition, and they instantly resume their fertility. Tame pigeons, sheep, mares, and numberless sorts of other animals, when they are stuffed to satiety, do not want to, or do not care to raise, or are incapable of raising others."

Mr. Doubleday's theory is that "the increase of population is connected with the food of the people." In other words,— and the truth of the deduction is evident to all observing men, —if the diet of women is composed of articles in which fatty constituents or carboniferous abound,—such as sugar, sweetmeats, nuts, and all the rich delicacies which are found on the tables of the rich and luxurious,—they will become plethoric, adipose, and *sterile*.

If, with this rich and unnatural diet, we find *lack of exercise*, an absence of outdoor life, and a condition of physical indolence, we have the two causes combined which as much if not more than all others conduce to render the female population of our large towns and cities unproductive and sterile. If we contrast this sterile class with the hard-working, hardy, plainly fed, and plainly clad portion of the population,—the Irish, German, and English,—we shall see at once the difference which exists between the two classes in the capacity for bearing children.

Hippocrates was well aware of these facts, for he says: "The want of fruitfulness arises from sedentary life, indulgence in riding in carriages, want of exercise, profuseness in living, fatness, and muscular laxness in the female sex."

Aristotle says: "The condition most favorable to procreation is a habit of body inured to labor."

Lord Bacon says: "Repletion is an enemy to generation."

Herbert Spencer, one of the profoundest writers and observers of the age, has the same opinions.

It is to be hoped that the renewed interest of the present generation in athletic outdoor exercises will have a favorable effect on the organizations of women, and in time undo the evil caused by a century of false fashion.

For a further consideration of this subject, so important as a cause of *sterility*, I refer the physician to that excellent paper on *The Law of Human Increase; or Population based on Physiology and Psychology;** by Dr. Nathan Allen, of Lowell, Massachusetts.

---

* Quarterly Journal of Psychological Medicine, April, 1868, vol. ii, p. 209.

CHAPTER IV.

II. PSYCHICAL CAUSES.

THAT there are causes of sterility which exist solely in the mental constitution of women, no close observer will deny. The annals of medicine and jurisprudence abound in cases of this character. Among the psychical causes of barrenness may be enumerated:

INCOMPATIBILITY.—There have been many instances known where, in spite of perfect development of all the organs of generation, together with a normal healthfulness of functions, in both husband and wife, conception did not occur. The reason why such marriages were unfruitful is said to be some antipathy or antagonism in the mental sphere. Gollman* mentions such instances, and calls them *relative* causes of sterility. He says:

"Such causes are, for example, antipathy or antagonism between the married parties; extreme differences of age, constitution, and temperament, as when a very old man is married to a very young wife, or *vice versa*, a very old woman to a very young husband; or when a cold and phlegmatic husband has a wife of a very ardent temperament. Such causes are quite frequent, although it is not always easy to find them out. There are instances of sterility of five, ten, or twenty years' standing, and when, after this period, conception took place in consequence of a change in the outward relation of the married partners."

Several cases have come under my own observation, where husband and wife have lived together for ten or twenty years without the occurrence of conception. These parties became divorced, each married again, and had children. The case

---

* Diseases of Urinary and Sexual Organs, p. 248.

of Napoleon and Josephine is in point. The Emperor had children by his second wife, and Josephine's first marriage was fruitful. "We frequently see trials of divorce," says Casanova, "reported in the newspapers, brought in by husbands against their wives for adultery, when the latter had been unfruitful during a number of years of matrimonial life, and became pregnant by other men. Infidelity in such cases was generally found to be the result of ill-treatment from their cruel husbands; consequently their wives could not possibly have had the least degree of love and affection, and therefore no power of reproduction during such behavior."

This last sentence embodies the doctrine taught by this singular author in his unique, but illogical work,* heretofore referred to.

FRIGIDITY.—Coldness, or an absence of sexual desire, was once supposed to be a cause of sterility; but the most eminent physiologists of this century do not teach that such a condition is a cause. Absence of sexual feeling may arise from some malformation of the external genitals, while at the same time the ovaries perform their functions properly. The consideration of this delicate question lies in the domain of psychology.

Treatment.—When no organic cause is discovered, the remedies are: *Kali brom.*, *Agnus castus*, *Conium*, or *Baryta* in the high attenuations; or *Phosphorus*, *Helonias*, *Cantharides*, *Sumbul*, *Moschus*, *Ergot*, or *Nux* in the low attenuations.

EROTISM.—Excess of sexual excitation or enjoyment in one or both, especially in hysterical and passionate women, is stated by some authors to be a cause of sterility. This, like the foregoing, is a subject which I do not propose to discuss in this work. I prefer to adhere principally to the more practical details of the pathological causes of sterility.

Treatment.—Erotism, or erotomania, is best removed by *Cantharis*, *Phosphorus*, *Nux vom.*, *Platina*, *Lilium*, *Origanum*, *Moschus*, *Sumbul*, *Cannabis ind.*, in the higher potencies; or, *Kali brom.*, *Lupulin*, *Camphor*, or *Ferrocyanuret of potassa* in the low attenuations, or massive doses.

---

* Contribution to Physiology.

CHAPTER V.

OVARIAN.

ABSENCE OF THE OVARIES.—Congenital absence of the ovaries has been observed in a few instances. The persons so abnormally constituted possessed some of the characteristics of woman, but were more masculine than feminine in their appearance. They may menstruate, but they cannot conceive.

IMPERFECT DEVELOPMENT.—This condition, which consists in a persistence of the fœtal condition of the ovaries, is not very rare. It may exist on one side only, though it generally affects both. As in the case of absence, a certain conclusion is not easy; and, as in that case, also, we draw a presumptive conclusion from want of development in the other organs of generation, absence of the usual signs of the menstrual crisis, and lack of general constitutional vigor and development.

Prof. Thomas* gives three cases of this condition which came under his own observation. Their ages were respectively 24, 28, and 30. Each showed a want of development of the uterus, vagina, and external organs of generation. The first two never menstruated, the latter very scantily.

Treatment.—In imperfect development the treatment should consist in administering the remedies homœopathic to the condition, namely: *Conium, Iodine, Chimaphila, Helonias, Ferrum, Phosphorus, Calc. phos., Baryta carb., Senecio,* and the *Hypophosphite of lime.*

The action of these medicines may be aided by careful and judicious irritation of the uterus by means of slippery elm tents introduced at or near the probable menstrual periods. Electricity has been successfully employed in such cases. It acts by arousing the dormant forces in those organs. One

---

* On Diseases of Women.

pole of a battery should be placed on the spine, and one over the ovaries; or one pole in contact with the cervix uteri, and one over the ovaries.

These means, together with proper bathing, exercise in the open air, a nutritious diet, etc., will be apt to arouse the undeveloped organs.

ATROPHY.—Aside from the normal atrophy which occurs at the change of life, the ovaries may become atrophied at any age. It may result from acute ovaritis, pelvic cellulitis, and peritonitis. If both ovaries are affected in this manner, sterility will of course be absolute.

Treatment.—If the atrophy be not too far advanced, it may be arrested by the same means recommended for undeveloped ovaries.

An excellent paper on this subject will be found in the *North American Journal of Homœopathy*, vol. ii, p. 192,* in which the treatment is considered quite extensively.

INFLAMMATION.—Acute inflammation of the ovaries may cause such an amount of suppuration and disorganization as to destroy forever their functions; or it may end in chronic inflammation, resulting in destruction of the capability of elaborating healthy ovules. If only one ovary be diseased sterility will not result.

Treatment.—The specific remedies for ovaritis are: *Aurum, Apis, Cantharis, Conium, Lachesis, Clematis, Pulsatilla, Rhododendron, Platina, Lilium, Thuja, Sabina,* and *Phytolacca.* In very severe cases with high fever, *Aconite, Veratrum vir., Gelseminum,* and *Belladonna* must be given.

DEGENERATION of the ovaries may result from acute or chronic inflammation. When the disease has reached this stage, no treatment is known that will restore their integrity; and if both ovaries are affected, permanent sterility must result.

---

* On "Atrophic Ovarian Amenorrhœa."

TUMORS of the ovary are not uncommon. They may be divided into: (1.) Those which are solid. (2.) Those which have cavities, more or less large, in their interior; dropsical tumors. (3.) Cancerous (scirrhous). The remedies recommended are: *Apis, Arnica, Belladonna, Conium, Graphites, Lachesis, Lycopodium, Zincum,* and *Lilium.* Some remarkable cures have recently been made by means of electricity. Ovarian tumors of large size have decreased notably, and in some instances apparently disappeared, under the use of *Chlorate of Potassa, Kali bromatum,* and *Chimaphila.* Dr. Guernsey has collected a number of cases supposed to be cured by homœopathic remedies (*Hahnemannian Monthly,* December, 1877).

DROPSY.—Ovarian dropsy is but a form of ovarian tumor; a local disease, and not dependent on disorder of the kidneys.

CHAPTER VI.

UTERINE.

Fallopian Tubes.

Any abnormal condition of the Fallopian tubes sufficient to prevent the passage of the ovum from the ovary to the cavity of the uterus is a cause of sterility. It matters not whether impregnation takes place in the ovary or the uterus, normal conception cannot occur unless the ovum attaches itself to the internal wall of the uterus.

Among the diseases of the Fallopian tubes, causing sterility, may be mentioned,

(*a.*) STRICTURE.—It is said by several European writers that stricture of the uterine end of the Fallopian tube may occur as a result of previous inflammation. The occurrence of a fœtus being developed in the tube seems to prove that such a condition may occur. Such an abnormality, however, cannot be demonstrated during life, unless some unusual dilatation or organic change in the os uteri coexist. "Theoretically," says Dr. Gardner, "there is little doubt of its existence; but practically there is an immense doubt of the possibility of recognizing it, save, perhaps, in some exceedingly rare instances."

Treatment.—The recommendation of some recent English writers, Tyler Smith, the originator, of passing a bougie into the os and through the cervix, and thence into the Fallopian tubes, and by a series of graduated instruments removing the stricture, is itself so seemingly impossible that it is not now seriously recommended. It has never been performed, and doubtless never will be. Gardner[*] gives cuts illustrative of the method of performing this theoretical operation.

---

[*] Sterility, pp. 156–7.

(*b.*) OBSTRUCTIONS may occur anywhere in the course of the tubes, but generally at the two extremities.

Tyler Smith says: "I have repeatedly observed that in cases where the Nabothian bodies are met with, small cysts are found in the course of the Fallopian tube, or at its fimbriated extremities. Inflammation of the tube may result in exudation into the tube, blocking up the canal, and rendering it impervious. Catarrhal affections of the tubes may cause obstructions of a more or less persistent character. Polypoid tumors may also fill up the canal of a tube."

Treatment necessarily obscure.

(*c.*) INFLAMMATION of the Fallopian tubes may be discovered during life, but its diagnosis from ovarian inflammation is extremely difficult. The tube itself may be the seat of inflammation, causing stricture or obstruction; or the fimbriated extremity may be affected, resulting in *adhesions* or *displacements*.

(*d.*) ABSENCE OR DEFECT OF THE UTERUS, in respect of development, will cause absolute sterility.

"*Complete* absence of the uterus has been doubted by many, because a more careful study of cases recorded as authentic, rendered the presence of the rudiments of a uterus presumable or evident; and partly because cases of absence of a uterus which are founded simply on examination of living women do not seem sufficiently conclusive. Notwithstanding this, it has been proved that the uterus may be entirely absent, and in such cases both the oviducts and ovaries, especially the former, may exist either in a rudimentary state, or may be also wanting."*

Burggraeve relates cases where, with absence of the uterus, both the ovaries were normally developed, with Graafian follicles found in them; but the menstrual flow and sickness were absent. Scanzoni speaks of extravasations of blood and formation of cysts, which cannot be imagined to take place

---

* Kolb, Pathological Anatomy of the Uterus.

without ovulation. The breasts are frequently well developed, the pelvis of the normal dimensions, and sexual desire not absent. In this anomaly, as well as the *rudimentary* uterus, described by Kolb, conception cannot occur. The absence of the menstrual flow is, in such cases, a valuable diagnostic symptom.

ATRESIA of the uterus may be congenital or acquired. Of the former, Kolb writes: *

"By *congenital* atresia we understand that arrest of development in which the external orifice, or, at the same time, a portion of the cervical canal, is either imperforate or entirely absent. As the extremest degree of this anomaly, we might consider that rudimentary form in which no cavity has been formed. However, the term atresia is only applied to the slighter anomalies just mentioned."

"The cervix of the uterus is, in rare instances, perfectly imperforate, and forms a slender cylindrical body, consisting of fibrous connective tissue, interspersed with muscular fibres. In such cases the vaginal portion of the cervix is either entirely absent or very imperfectly developed and small, and the vagina is always rudimentary. In other cases the atresia is limited to the external orifice; the occlusion being also effected by muscular and connective tissue. In such, the vaginal portion is generally small, but in some instances it is found normally developed. Finally, we meet with cases in which the vaginal portion and the cervix are perfectly permeable, but the atresia is formed by the mucous membrane of the vagina passing over and occluding the orifice of the vaginal portion."

Congenital atresia, in all its forms, is a very rare anomaly. It may exist unperceived up to the years of puberty, and acquire pathological importance only at the commencement of menstruation. The menstrual fluid accumulates in the uterine cavity, from which it cannot escape; and, consequently, distending the uterus, it causes that condition termed

---
* Pathological Anatomy of the Uterus, p. 32.

hæmatometra, or *retention* of the menses. All these forms of atresia cause sterility. A uterus affected with atresia cannot be impregnated, yet after a successful operation, not only conception, but maturity of the fœtus and a normal delivery have been known to take place.

Acquired atresia, occlusions, or contractions of the uterine cavity take place usually at its *orifices;* and as regards frequency, those at the internal orifice are more frequently observed than those of the external, whilst throughout the remainder of the uterine cavity, stricture or occlusion is rarely found. Sometimes the entire uterine canal is diminished by concentric hypertrophy or atrophy.

Stricture of the uterine cavity may be caused by *annular contractions at a certain point;* by the *external pressure of fibroid and other tumors;* by *partial or complete flexions* of the uterus; thickening of the mucous membrane from catarrh; enlargement of the follicles, or from the growth of granulating tissues resulting in a constricting cicatrix. The abuse of caustics, such as *Nitrate of silver*, *Chromic acid*, and *Potassa fusa*, in the treatment of ulceration and inflammation of the cervix, is doubtless a frequent cause of occlusion. One such case came under my observation, and required an operation for its removal. In the treatment of ulceration during pregnancy, proper caution must be used or the edges of the os may unite. This has been known to occur when no local treatment was resorted to.

Sterility may result from nearly every form of acquired atresia.

Treatment.—For complete *congenital* atresia of the uterine cavity, no surgical treatment can be of avail. In those forms of partial atresia, puncture or incision through the occluded portions, with the usual precautions against re-unions, will be attended with success.

(For the purpose of diagnosing contractions of the uterine cavity, or atrophy of the uterus, *Fitch's measuring-sound is best.*)

Fig. 1.

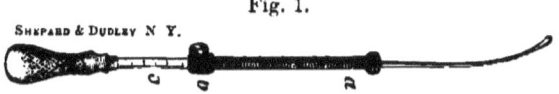

Acquired occlusions or contractions, if confined to the canal of the cervix or its inner or outer orifices, may be successfully treated by the same surgical means. If from *flexions* of the uterus, a proper pessary and tents will remove the stricture. (See Flexions and Versions of the Uterus.) Tumors may sometimes be removed by surgical operations, or the persevering use of such remedies as *Conium, Hydrastia*, and *Iodide of arsenic*. Simple occlusions from the adhesion of the lips of the os can readily be removed by the uterine sound, or in severe cases by the bistoury. Lacerations of the os and cervix should be watched for after miscarriage and delivery. In closing up they often completely occlude the os, and it is not ascertained until after the menses occur at the close of lactation. Instances have been reported where occlusion of the cervical canal has resulted from the use of instruments used for the purpose of criminal abortion. These causes should not be forgotten, and their results guarded against. If such a condition threatens to occur from any of the above causes, the os and cervical canal should be kept open by means of tents and bougies until the danger is past.

IMPERFECTLY DEVELOPED UTERUS.—I have known of several cases of sterility in otherwise healthy women from this cause. Such women are nearly always inclined to obesity.

Dr. Simpson* says he has seen a considerable number of cases of short or imperfectly developed uteri in the living subject. The imperfect development of the organ was ascertained, on examination by the finger, by the small and atrophied cervix uteri, and by actual measurement of the length of the cavity by the uterine sound. Instead of being two and a half inches in length, the cavity in such cases was only two inches, or more frequently only one and a half or one inch long. The subjects of this imperfect development were often well made and well formed in other respects. But the malformation led to various functional defects, especially to *amenorrhœa* and *sterility*. The amenorrhœa was usually

---

* Obstetric Memoirs, vol. i, p. 258.

persistent, and when a patient applied for medical relief who was already twenty-five or thirty years of age, this malformation would in a large proportion of cases be found to be the organic cause. In some such cases of amenorrhœa, there was great vascularity of the face, and occasionally a most unconquerable form of acne. He has seen in some of these instances the wearing of an intrauterine galvanic or zinc and copper pessary followed by the best results, and even occasionally by the cure of the amenorrhœa. The uterus developed itself around such a foreign body when it filled its cavity as it did around a fibrous tumor or an ovum.

Dr. Storer, in his note to this article, remarks that he has often seen galvanic pessaries used by Dr. S. in ordinary cases of amenorrhœa which had resisted all the usual means of treatment, and with perfect success.

Imperfect development will not only cause amenorrhœa, with sterility, but it causes that kind of difficult menstruation in which the flow is *very scanty* and sometimes *delaying*. This condition is nearly always attended with *sterility*. I do not think the imperfect development is always congenital, but oftener arises from a sudden check to the first menstruation. I have now two such cases under treatment. In both a severe cold checked the first or second menses, which began profusely enough, but ever since that date they have been scanty and delaying. Each has been married thirty years, and never have been impregnated.

Treatment.—No medicinal agent is of any service after the condition has existed a few months. If *Pulsatilla, Senecio,* or some remedy appropriate to the sudden suppression were given immediately, the normal development would not have been arrested.

ATROPHY of the uterus is an affection of mature age, and generally commences simultaneously with puerperal involution, or it must be considered as marasmic degeneration of the organ. In chlorotic women also, a sort of atrophy of the uterus is sometimes met with, generally complicated with displacements and derangements of menstruation. Besides these causes, atrophy may result from pressure, or be due to mechanical causes. As regards the extent of the affection, we may distinguish between general and partial atrophy. The latter affects either the body, fundus, cervix, or vaginal portion of the uterus. As regards the *cavity* of the uterus, a distinction has been made between *concentric* atrophy, with diminution, and *excentric* atrophy, with dilatation of the cavity. (Kolb.)

Concentric atrophy would doubtless be a cause of *sterility;* but it is doubtful if atrophy with dilatation would be inimical to conception, although it might act as a cause of miscarriage. Concentric atrophy generally occurs as a sequence to the puerperal state, and is probably one of the causes of that form of sterility which occurs after the woman has given birth to two or more children, and is often connected with atrophy of the ovaries.

SENILE atrophy, occurring as it does after the climacteric age, does not belong to the causes of sterility. Scanzoni mentions as a cause of atrophy, imperfect innervation of the pelvic organs consequent upon paralytic conditions of the system (paralysis of the lower half of the body, followed by amenorrhœa), and of which he observed several cases.

Treatment.—If the atrophy is from excess of puerperal subinvolution or from a psoric miasm : *Iodine*, if there is a general wasting of the body with great weakness of the joints.

Baryta carb., if there is an appearance of premature old age, diminution of sexual desire, and the uterus feels hard and is diminished in size; menses scanty or suspended.

Conium, if there has been swelling and soreness of the mammæ, followed by wasting, with induration and atrophy of the cervix uteri; sexual feeling abolished; menses scanty or wanting.

Chimaphila, if the breasts have become atrophied and flabby, *without* previous soreness; the skin is dry and hard; and the urine deposits a profuse white sediment, like mucus and chalk.

Graphites, Sepia, and *Kali bromatum* are indicated if the menses are *very scanty.*

If it is caused or attended by chlorosis:

Iodide of Iron, if with the *Iodine* symptoms there is waxen complexion, bloated feet, etc. *Ferrum*, if the chlorosis is due to deficiency in the red corpuscles of the blood.

Helonias, if there is chlorosis with dropsy or diabetes.

Bromine is an excellent stimulant of the ultimate cells of the ovaries and uterus, and may avert atrophy of either of these organs.

In atrophy from imperfect innervation, with paralysis of the lower half of the body, the remedies are: *Nux vomica*, *Conium*, *Secale*, *Ustilago*, and *Phosphorus*, or electricity.

Cervical Stenosis.

OBSTRUCTIONS from stricture of the os and of the canal of the cervix uteri, is of frequent occurrence, and often overlooked by ordinary practitioners. On the other hand, surgeons who are always looking for stricture of the cervix often mistake the effects of the various *flexions* of the uterus for the former condition.

Dr. Simpson has an excellent paper on the surgical treatment of this abnormality. He recommends incision, and mentions several cases of *sterility* from this cause cured by the operation.

Dr. Mackintosh recommends dilatation by long, straight bougies of different sizes.

Dr. Sims prefers the bilateral incision of the cervix.

Dr. T. G. Comstock, of St. Louis, writes: "One case of sterility I treated successfully by dilating the cervical canal by means of the uterine probe, Atlee's dilator, sponge-tents, and Simpson's graduated gutta-percha plugs. This operation, made in another case under treatment for a year past, has entirely failed, although the stricture has been overcome. For the last six months I have been experimenting in cases of dysmenorrhœa, from a narrowing of the cervical canal, by relieving the stricture by incision; that is, dividing it, making the bilateral section of the os and cervix uteri, which is the operation proposed and performed successfully by Drs. Sims and Simpson.

"This operation I consider preferable to dilatation, and the results of this practice I propose to make public at some future period. I am acquainted with a physician who claims to have performed this operation nearly fifty times, without any accident or untoward results, and to have cured many cases of sterility by it.

"Whether this operation is unwarrantable or not, it is certainly worthy the attention of gynæcologists and surgeons of our school."

The July (1877) number of the *American Journal of Obstetrics* contains a paper by Dr. Pallen, entitled, "Incision and Division of the Cervix Uteri for Dysmenorrhœa and Sterility." In that paper are published a number of questions addressed to Dr. J. Marion Sims, and relating to this operation and its results. Dr. Sims states that he has for several years operated with the instrument figured in his work on *Uterine Surgery*, and which has since been greatly improved. With it he makes a bilateral incision up to the cavity of the uterus, then dilating the canal by a *dilator*, and introducing a plug of hard rubber to keep it from contracting, and retaining the plug by means of styptic cotton. This dressing is allowed to remain from two to six days. Great care should be taken *not to incise the cervix too deeply*. The incision should begin at the inner os, which need not be opened larger than *one-fourth of an inch*. From this point the incision can be carried deeper as the *outer* os is approached, making the canal funnel-shaped.

Dr. Sims says he considers "the sponge and laminaria tents more dangerous." He states that he has operated nearly a thousand times and only lost two patients by it, who died of peritonitis. He pronounces the use of tents, for the purpose of enlarging and lengthening the cervical canal, as *utterly useless* for the permanent cure of dysmenorrhœa or sterility. I have found it so in my practice, for although it may relieve the dysmenorrhœa for two or three months, the stricture almost invariably returns. Incidentally, Dr. Sims mentions the impropriety of treating any case of supposed sterility without first ascertaining if the woman is really at fault, or whether the fault lies with the husband. He says he has met with many cases where the woman has been treated uselessly for years by various physicians, when at last the microscope demonstrated either that the spermatozoa were killed in the cervical canal by the abnormal secretions of the passages, or that they were dead or absent from the seminal fluids. I have myself met with very many similar cases, and I would advise every physician to examine the seminal fluid before and after it has been placed in the generative passage of the woman. To omit this is to "work in the dark," and allow the blame to rest where it does not belong.

There are several instruments for division of the cervix. Simpson's hysterotome, invented in 1843, was the first one used for this operation, and by it that distinguished surgeon gained a world-wide celebrity. It is single-bladed, the blade concealed, and cuts one side at a time. It is still used very successfully by many gynæcologists. There are several double hysterotomes, with two blades, cutting both sides at once. Of these, Drs. Greenhalgh's, Stohlman's, and Peaslee's are the most popular. Peaslee's instrument is perfectly safe, for it can be set so as not to cut too deeply. I have operated several times with his hysterotome, and afterwards introducing a plug of slippery elm two inches long and one-fourth of an inch in diameter, retaining it in place by a small tampon of cotton wet in a solution of persulphate of iron, introduced just within the lips of the os.

This is safer than Sim's formidable operation, an operation which, however safe in the hands of such a master surgeon as Sims, is one that the average practitioner should not attempt. Moreover, Sim's operation cannot be performed without using his speculum, which requires an assistant, while the operation with Peaslee's hysterotome can be performed unaided and through any ordinary speculum with movable or expanding blades. I offer no apology for here introducing Dr. Peaslee's famous paper on this subject, as its importance cannot be overestimated. He says:—

"If 'meddlesome midwifery is bad,' meddlesome surgery is not less so; and that form of it which attacks the uterus is the worst of all—since it not only injures the patient herself, but also compromises, and perhaps destroys, her prospects of offspring. Trachelotomy, or cutting the cervix uteri, has been of late so indiscriminately, and often so unnecessarily performed, as to suggest this general remark. And though perhaps somewhat less common, and with some operators less severe, than five years ago, a reference to the most authoritative treatises on gynæcology shows that it has not yet reached its legitimate limitation, even among gynæcologists; while many general practitioners operate as frequently and as blindly as ever—it being so facile a procedure that nobody hesitates to attempt it.

But Trachelotomy, in some form, must continue to be recognized in the treatment of many cases of dysmenorrhœa and of sterility, when depending on stenosis of the cervical canal; and the least hazardous, if equally curative, should be preferred. I propose to consider its usual methods and their uses, abuses, and actual value; and also to explain a new method, which, I maintain, includes all that is valuable in them and still more, without their objectionable characteristics.

The two authorative methods of Trachelotomy, hitherto practised, with some modifications hereinafter to be specified, are:—

I. By Simpson's Metrotome, or some modification of it—deep incision of the cervical canal.

II. By the scissors—discission of the cervix, or Sims' operation.

To these I add a third method, first suggested by myself, and which I will designate as "superficial trachelotomy."

Each of these methods will be separately considered.

I. DEEP INCISION OF THE CERVIX UTERI. SIMPSON'S OPERATION.

Prof. Simpson, of Edinburgh, maintaining that the stenosis producing dysmenorrhœa and sterility exists usually at the internal, and not at the external os uteri, in 1844 devised his metrotome for overcoming the constriction. It is shown by Fig. 2, and is too well-known to require any special description. I improved it, I think, some fifteen years since, by lengthening the sheath to the extent of three-quarters of an inch beyond the blade, so that the full strength of the fingers can be brought to bear upon the blade without displacing the sheath, in case the uterine tissue requires to be divided without the application of traction at the same time. Subsequently, Dr. Greenhalgh, of London, proposed a two-bladed instrument, it being merely a double, as Dr. Simpson's was a single "*bistourie cachée.*" This divided the cervix symmetrically, or very nearly so, as Dr. Simpson's instrument did *not*, except by chance—and, generally, more extensively than the latter was intended to do.

Figure 2.

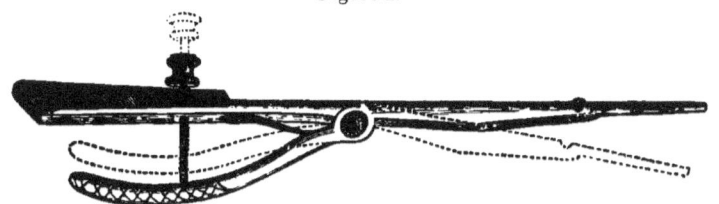

Simpson's Metrotome.

I might also mention several other metrotomes devised in this and in other countries, were they of any special importance to my present purpose. These are almost all double-bladed and act on the same principle as Dr. Greenhalgh's instrument, which soon became generally preferred to Simpson's. I therefore speak of its action especially in what remains to be said under this head; there being, in fact, no practical difference between its effects and those of Simpson's metrotome, when freely used, as by its originator himself.

Advocated by so able a defender, Simpson's operation soon became quite common in Great Britain and this country, though it was not accepted on the Continent till 1860. Its range of application also became extended. For while, rationally, it was at first invoked only in cases of stenosis of the cervical canal, including of course the two ora, it ere long became common enough in cases in which no obstruction at all had existed, and as a mere matter, as it were, of fashion. During a sojourn abroad in 1866, I witnessed a number of operations which I could place only in this category; and several of the same kind have I since seen at home. It seemed to be generally assumed that the uterus is an organ quite indifferent to cutting and hacking, and that the deep incision would, at any rate, do no harm. Indeed it was known that Dr. Simpson had often performed the operation at his consulting-rooms, and afterwards sent his patients home in a cab; and I cannot learn, up to the present time, that he ever reported an adverse case in his experience. It is, however, stated, on unquestionable authority, that some of his patients died in consequence of the operation, and others narrowly escaped death; and it is proper that I here definitely specify its immediate and remote effects. Of course, it is to be recollected that I am speaking of deep incision of the cervix, as performed by means of Greenhalgh's metrotome, or of Simpson's, in a bold hand like that of its originator. And I have to consider:

1. The change in the shape, size, and relations of the whole uterine cavity by the deep incision.

2. Its immediate dangers.

3. Its remote effects.

1. *The change in the shape, size, and relations* of the cervical canal, produced by the deep incision, are shown by Fig. 3,

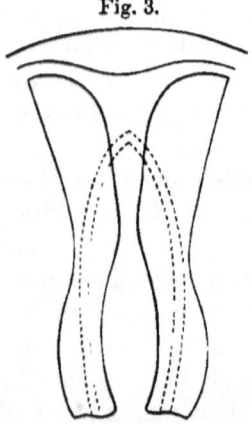

Fig. 3.

Large and small cut by Greenhalgh's Metrotome.—HEWITT.

where the normal shape and size of the canal are seen; while the dotted lines outside of it show the "smaller and the larger incision," as Dr. Hewitt calls them, made by Dr. Greenhalgh's instrument.

Peaslee denounces the operation by means of Simpson's or Greenhalgh's metrotome as dangerous and useless, for the following reasons:—

"It is seen that the internal os after these incisions is somewhat more than three times its normal width, while the rest of the canal is increased in nearly as great proportion; and that the whole cavity of the uterus, cervix and corpus together, no longer retains its normal form as shown by Fig. 8; but resembles an erect, wide-necked, flattened flask without a bottom, Fig. 11. We shall see whether such a cavity can be depended upon, as a receiver or as a retainer, farther on. This operation ignores the importance of the normal relations of these cavities, even more than would one, were such a procedure possible, which should permanently dilate the urethra to the size of the small intestine. I fully assent to Dr. Sims' criticism, that this operation cuts altogether too extensively; an objection which will, however, be seen to apply as truly to his own.

2. *The immediate dangers* of so deep a division of the cervix are, a profuse and sometimes even a fatal, hemorrhage, pelvic cellulitis, and septic peritonitis, which is almost always fatal. It is known that these results occurred to some of the patients operated on by Dr. Simpson at his consulting rooms. There is also a risk of cutting through the cervix into the peritoneal cavity.

Such effects can, however, surprise no one who is aware of the extent of the lesion produced. Indeed, it is surprising, rather, that they are not more frequent than they are actually found to be. In the operation by Greenhalgh's metrotome the walls of the cervix are cut more than half through on both sides to a considerable extent by the lesser incision with the two-bladed metrotome, while, by the greater, should there be a slight inclination of the uterus or of the instrument to either side, or a slight thinning of the uterine walls—an opening would be made into the peritoneal cavity. This has actually occurred. The entire cervix may also be split completely through, as in a case mentioned by Dr. Sims (p. 171).

A free hemorrhage, at least, is inevitable in such circumstances, and even the lesser incision may divide one or more arteries on the level of the internal os, as shown in Figs. 4 and 5; where it is seen that, in the second preparation, there is an

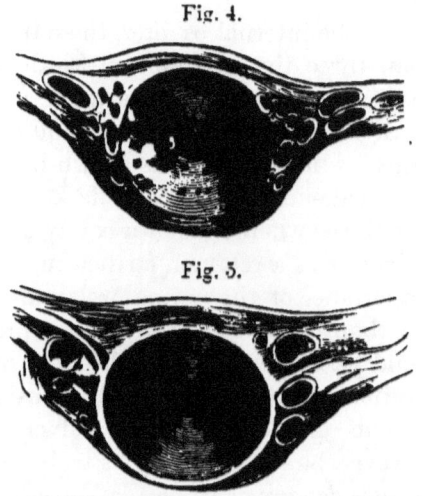

Fig. 4.

Fig. 5.

Section of uterus made at os internum— ad nat.

Showing the normal size of the os internum, the circular disposition of the fibres around it, and the blood-vessels in proximity.—BARNES, p. 207.

artery within one-eighth of an inch, and in the other, three arteries within one line, of that opening. The lesser incision cuts to the depth of at least three-sixteenths of an inch, and the greater to more than one-quarter. If sufficient for all practical purposes, it is therefore of the greatest importance to restrict the depth of the incision of the internal os within the limits of the arterial distribution, and which I shall show may be done.

The dangers of pelvic cellulitis and septic peritonitis are referrable to the fact that the medullary portion (Savage) of the cervix is laid open, and that thus perfect facility is afforded for the absorption of septic matters. This also may be avoided.

Hence Dr. Barnes [*] regards "incision of the internal os as

---

[*] Dr. Barnes' objection to Dr. Greenhalgh's instrument, because it acts automatically, is not, I think, well taken. If an instrument acts automatically, but

being attended by great danger." First, there is profuse, even 'furious' bleeding; next, from the gaping of the divided veins, and the injury of the tissues in which they run, there is liability to pelvic inflammation and septicæmia. These are no theoretical dangers. Many cases, some fatal, are well known" (p. 206). He says that "an incision even one-fourth of an inch deep will be very liable to divide some of the vessels;" and therefore, though he still sometimes uses Simpson's metrotome, he cuts only from below the internal os to the external.

I am obliged to speak thus generally of the dangers of Simpson's operation, since no statistics have ever been published, as should have been done, especially by those who have operated most frequently.

From the facts I have given, it may be inferred that no amount of experience in this operation will prove a safeguard against its dangers, and the following case illustrates this, as well as the unpardonable carelessness which great familiarity with an operation sometimes engenders:

A lady, 28 years of age, who had been married eight years without ever having been pregnant, applied to me several years ago, to remove the cause of sterility. She had no dysmenorrhœa, no uterine displacement, no stenosis of the cervical canal—nor, indeed, any uterine symptom at all, excepting a slight leucorrhœa, from congestion of the endometrium. This having been cured, she soon after left, with her husband, on a short summer trip to Great Britian and the Continent, I having advised her, if any uterine symptoms returned, to get the advice, while in Edinburgh, of the then most distinguished gynæcologist there. No uterine symptoms did return; but while in that city for two or three days, she decided not to lose such an opportunity to obtain his advice respecting the sterility, and sent for him. After a rapid examination, he

---

always within certain perfectly well-known and safe limits, thus enabling the skilful surgeon to succeed, and securing the bungling operator against an accident, it is not to be rejected merely because automatic. But if an automatic instrument acts at random, or always beyond certain limits of safety, causing both the scientific and the ignorant operator equally to do harm, it is certainly very objectionable. This I hold to be the *real* objection to Greenhalgh's instrument.

remarked that a very slight operation was required, and at once introduced his metrotome, and incised the cervical canal, and left the room within about three minutes afterwards. The husband left about five minutes after the physician, and did not return for an hour, when he found his wife had fainted from loss of blood, which had saturated the bed, and escaped upon the floor. Rushing to the doctor's residence, he found the latter had gone five miles out of the city, to the wedding of one of his assistants; and going next for another assistant, he also was found to have gone thither. The nearest physician was then called in, who found there was no time to be lost, and arrested the bleeding by continued pressure, until the operator could be brought back to the city. Returning, he remained with her for the next twelve hours. Her life was barely saved; and, after passing the summer in that apartment, she had only recovered sufficiently to be able to travel; and the time for the journey to the Continent being exhausted, she returned directly home. She required more than a year for the recovery of her strength and color, and now at the end of seven years, still remains childless. The surgeon had performed the operation, probably, more times than any one else in Europe, and must have previously found it to be a treacherous procedure.

3. Having found the immediate effects of this operation thus undesirable and even dangerous, I now proceed to consider *its more remote results*, both curative and otherwise.

As a remedy for dysmenorrhœa and sterility, when depending on stenosis of the cervical canal, Simpson's operation usually succeeds, temporarily at least, with the former; and as generally fails with the latter. As the contents of a bottle without a bottom can have no difficulty in escaping from it, so the menstrual fluid should have none in leaving such a uterine cavity as that shown by Fig. 11. And if pain still attends menstruation, it is, of course due to some other cause than stenosis. The incision, however, not seldom gradually closes up, in spite of the surgeon's intentions, and the relief proves to be temporary; and sometimes the cicatrix, continuing to contract, finally reproduces the dysmenorrhœa in a severer form than existed at first.

On the other hand it can scarcely be expected that such an enlarged open cervical canal, as Fig. 11 represents, can exert any active influence in favor of conception, or retain the spermatic fluid, if by chance entering it. Besides, if pregnancy should actually supervene, in spite of such conditions, the ovum would probably escape from the uterine cavity prematurely. And these expectations are confirmed by observation. Conception but rarely follows the operation, as performed by Dr. Simpson; and when it does, abortion is very likely to ensue. Dr. Gream* reported a case of this kind; Chrobak has had several cases, and I have myself known of six. Scanzoni admits that dysmenorrhœa is frequently relieved by this operation, but objects that sterility persists notwithstanding. Hegar and Kaltenbach had fair success in the former, but their results in the latter were "less brilliant." And Barnes remarks that "the cure of sterility is not nearly so frequent as the cure of dysmenorrhœa" (pp. 212, 213). I think the main facts on this point may be summarized as follows:

1. If the incisions close up, there is for a time an increased chance of conception; but the progressive induration and deformity of the external os—since nature generally effects the closure very awkwardly—finally increase rather than diminish its original improbability. If the incision remain entirely unclosed, the sterility is generally confirmed.

2. The further the operation stops short of the deep incision of Simpson, the better the prospect of curing the sterility. Hence, in some hands, it not very seldom succeeds, though performed by Simpson's or Greenhalgh's metrotome, simply because the cervical tissue is not divided deeply, as was done by the former. A distinguished obstetrician of this city informs me that he cured his first two patients of sterility by Simpson's metrotome; but has very seldom succeeded since, for the reason, I suppose, that he divided the cervical canal but very slightly at first, and became bolder from experience. Whatever of success Chrobak has had, in the treatment of sterility by the incision of the cervix, I attribute to his modi-

---

* Dr. Sims, p. 170. From Lancet, April 8, 1865.

fications of Simpson's method. He incised the internal os but three times in 250 cases. He, indeed, varied his operations, and also resorted to other treatment, to such an extent in different cases, that his statistics are of no value for my present purpose, except so far as they recognize the decided tendency to abortion, if pregnancy afterwards ensues. There cannot, I think, be a reasonable doubt that Simpson's operation, performed on a woman in perfect health, would almost certainly render her sterile, unless the incisions closed up.

3. Sterility is more likely to be cured by the deep incision, as Dr. Barnes remarks, "the younger the patient and the more recent the stenosis," and before its complications are developed. I will also add that more prompt and complete closure is likely to occur in such patients.

The last remote effect of this operation which I shall mention, is the eversion of the labia uteri, in case the vaginal portion has been cut through on both sides. I agree with Dr. Sims that this is a very grave objection, though his own operation will be seen to produce the same effect far more certainly and frequently.

Finally, in specifying the *uses* of Simpson's incision of the cervix, I cannot recommend it in the treatment of stenotic dysmenorrhœa or sterility. It frequently cures the former, and the latter very seldom. But it is unnecesarily severe and dangerous; and a safer operation may and should be substituted. Of the two metrotomes, I would not recommend Greenhalgh's in any uterine condition which occurs to me; but Simpson's will be found a very useful instrument for incising uterine fibroids projecting into the cervical canal or the uterine cavity, and still covered by the uterine mucous membrane."

II. Discission of the Cervix. Sims' Operation.

Of discission of the cervix uteri—Sims' operation, he says:—"This method of Trachelotomy was first practised by Dr. Sims in January, 1857. His operation consists:

1. Of a complete severing, or discission* of the whole of the

---

* From *discindo*, to sever, to cut apart.

vaginal portion of the cervix, up to its vaginal attachment on both sides, by scissors; and

2. An incision of the whole canal above, and including the os internum on both sides, by a narrow, razor-pointed knife.

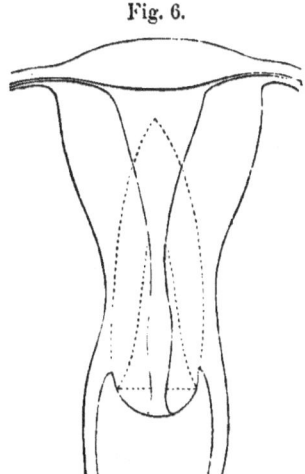

Fig. 6.

Discission of the Cervix.—SIMS, p. 172.

Fig. 10 shows the condition of the parts after it. Since the former is taken from Dr. Sims' work on "Uterine Surgery," it may be regarded as expressing his average intention in respect to the extent of the second step in the operation; for it is impossible for any one to know precisely how much he has cut into the canal, or to certainly cut the two sides precisely alike. From the tendency to hemorrhage after this operation, I think the cuts at the internal os are usually deeper than here represented. The cervix is also only three-fourths of its normal size.

Dr. Sims claims that his operation does not divide the entire cervix to so great an extent as the metrotome of Greenhalgh. But I shall show that the division is still much in excess of what is required, while it is attended by as great a change in the uterine cavity, and the same dangers in greater degree; and is followed by as undesirable remote results.

1. It is shown by Dr. Sims' diagram (Fig. 6) that the first

step in his operation practically shortens the cervical canal to an amount precisely equal to the length of the vaginal portion, or, on an average, nearly half an inch; for its discission on both sides practically annihilates it as completely as if it had been entirely removed by amputation. The relative length and shape of the whole uterine cavity, including the cervical canal, after this operation, as compared with its normal condition, is shown by Figs. 10 and 8. It has assumed the form of a flattened hour-glass, and the cervical canal is no longer fusiform, has lost nearly one-third of its length, and become as capacious as the cavity of the corpus.

2. The *dangers* of the operation are the same, but considerably exaggerated, as those of Simpson's operation, viz., hemorrhage, pelvic cellulitis, and septic peritonitis.

The danger of hemorrhage from the arteries around the os internum (Figs. 4 and 5) is not materially greater or less than in Simpson's operation. A bold or a careless operator will cut as far with the knife as with the metrotome, and still more at random and less symmetrically; while a timid cutter will probably go less deep with the knife without less probability of dividing the arteries. But, in another respect, the danger of hemorrhage is much greater in Sims' operation, viz., from the fact that the whole vaginal portion is completely severed, on both sides, up to the vaginal attachment. An alarming bleeding is also liable to occur suddenly at any time within the four or five days after the operation, though it was not very profuse during it, and special precautions should be taken both to prevent it, and to arrest it if it takes place. It is usually arrested at the time of the operation by the application within the incision, of cotton dipped in the persulphate of iron, and of the vaginal tampon. But a surgeon who had performed this operation very many times, assured me that he never undertook it unless he could have his patient so situated that she could be visited at the shortest notice by himself or a competent assistant, for the next four or five days. But in spite of these precautions, fatal hemorrhage has ensued.

The risk of pelvic cellulitis and of septic peritonitis is also greater in discission of the cervix than in Simpson's operation.

For, since the cut extends *through* the medullary portion of the cervix on both sides up to the vaginal junction, in addition to the incision in the canal above, a greater surface and facility is afforded for the absorption of any septic agent.

I fully agree, therefore, with Dr. Thomas, that "had all the fatal cases which have occurred in consequence of this operation been published, as they should have been, the list would be a startling one" (p. 414). He had known of five cases, and had rumors of others. I could myself add as many more.

It would seem hardly necessary to state that an operation liable to jeopardize a patient in the ways just specified should never be performed without an imperative necessity. Dr. Skinner, indeed, maintains that the vaginal portion should never be cut through.* This, however, like the preceding procedure, has been but too often practised apparently as a matter of fashion, or of the force of habit; for I can give no more charitable explanation of several cases which have fallen under my observation. In one instance, no reason had been assigned for the discission. It could not have been done, I think, for dysmenorrhœa, for the patient had passed the menopause at least ten years before; and the age of 55 to 60 years would also seem to exonerate her, also a maiden, from an operation for sterility. In another instance, the patient had a short time previously been under my care, and I knew she had no stenosis of the cervical canal, nor dysmenorrhœa, from any cause. And the operation was hardly demanded for sterility, since she had been married but four or five months. I ascribed it to the force of habit, or possibly to a reckless "*besoin d'opérer*" which sometimes possesses a mere surgeon.

3. The *remote effects*, curative or otherwise, of discission of the cervix are in general the same as those of Simpson's operation, already specified. That is, it generally relieves, and often cures, dysmenorrhœa, when depending on stenosis; while as generally it does not cure sterility, and does predispose to abortion, if pregnancy occurs. Indeed, in these last two respects, it is more objectionable than Simpson's method. For:

1. The severance of the cervix on both sides not only at

---

* Liverpool Medical and Surgical Reports, 1865.

once destroys all contractile force in favor of conception, as completely as if it had been entirely removed; but

2. The two pendulous flaps also act as valves to prevent the entrance of the spermatic fluid into the cervical canal. Besides, the retentive power of the proper uterine cavity is diminished by the practical ablation of the vaginal portion, and the change in its form, as just explained.

I should therefore apply to this operation also, as far as its value in the treatment of sterility is concerned, the three propositions already applied to Simpson's incision (p. 361).

Sometimes, however, the flaps become everted, and present the same appearance as if the cervix had been ruptured to the vaginal attachment in parturition; which latter condition, like the high amputation of the cervix, is well known to produce sterility in most cases, and to predispose to miscarriage, should pregnancy, notwithstanding, occur. This eversion mentioned by Dr. Sims as an objection to Simpson's operation, occurs far more frequently in his own; since, in the latter, the entire cervix is *always intentionally* severed, while in the former it is only seldom and accidentally so. All gynæcologists at the present day recognize the importance of the re-closure of the cervix ruptured in parturition. I maintain that the same should be done here; and the sooner after the discission the better. In other words, if a surgeon finds he has severed the cervix on both sides to the vaginal junction, in his treatment of dysmenorrhœa and sterility, he should *at once close it up again* by the proper operation. After a few months the flaps become atrophied to such a degree that it will be difficult, and perhaps impossible, to restore the external os and the cervical canal to their normal shape and dimensions.

I have already alluded to the tendency to closure of the incisions after this operation; and which is, comparatively at least, a fortunate event, as increasing temporarily, if not permanently, the chances of conception. But nature generally accomplishes this in a very imperfect manner. The two incisions are scarcely ever completely closed; and sometimes the attempt is made only on one side, in which case the contractility of the vaginal portion is still completely lost. If the two incisions are unequally closed, the form of the external os is, of course, abnormal; and equally as if no closure had occurred, is unfavorable for conception. And, finally, if both incisions are quite closed, just as if only partially so, the cicatricial tissue becomes indurated and contracted; so that in the end, a return of the stenosis and of the dysmenorrhœa, and in an aggravated form, may be the consequence.

Such being the merits of discission in the treatment of dysmenorrhœa, I should next speak of it as a remedy for anteflexion."

From the preceding facts, Peaslee deduces the following conclusions:

"1. The deep incision of the cervix throughout, and complete bilateral discission of the vaginal portion with deep incision above, are alike frequently attended by certain immediate dangers, and, not seldom, productive of certain serious remote consequences; viz., profuse and sometimes fatal hemorrhage, pelvic cellulitis, septic peritonitis (usually fatal), sterility, (if not previously existing), and a tendency to miscarriage.

2. Those risks and effects are all due to the extensive division of the walls of the cervix, and to the consequent enlargement of the cervical canal; and the sole compensation for all of them which can be calculated upon, is the relief, and very often the cure of stenotic dysmenorrhœa.

It therefore becomes a question of very great practical importance whether the amount of cutting may not be so far diminished as to avoid all these risks, and at the same time be sufficient for the cure of stenotic sterility and dysmenorrhœa. But another inquiry, antecedent to this, is:

How large a calibre of the cervical canal is actually required for the relief of these two conditions?—and a reply sufficiently definite for all practical purposes is not so difficult as might appear.

In the *imparous* woman, the narrowest point of the cervical canal, viz., the internal os, is, when opened by the passage of the menstrual fluid, an ellipse, whose conjugate and transverse diameters average respectively $\frac{1}{6}$ and $\frac{1}{8}$ of an inch; its area corresponding very nearly with that of a circle $\frac{1}{7}$ in. in diameter.* The external os, also elliptical when moderately dilated, has diameters averaging ¼ and ⅛ of an inch. It thus has an area exactly twice that of the internal os, and equalling that of a circle ⅙ inch in diameter.† The larger size of the external os doubtless has a special reference to conception, and favors the entrance of the spermatic fluid into the cervical canal. It has no special influence against dysmenorrhœa; since the menstrual fluid, after having passed through the internal os into the cervical canal, would pass just as easily from the latter through an opening of the same dimensions into the vagina. Hence we not very seldom see imparous women with the external os no larger than a "pin-hole," and who nevertheless do not suffer from dysmenorrhœa, though, as a rule they are sterile. But if the lining membrane of the canal becomes thicker, from congestion or some other cause, such patients suffer at once from stenosis at the external os.

In the *parous* woman, the size of the external os varies within quite extensive limits, since it is exposed to so many of the accidents of parturition; while the internal os is more nearly uniform.

I have deemed it desirable to ascertain the lowest average diameter of the two ora uteri in parous women, who are neither sterile nor have dysmenorrhœa, as a rational standard for determining the extent of incision actually required for the removal of these two conditions, when stenotic. And after a good deal of observation in this direction, I find that the inner os presents nearly twice the area of that of the imparous woman: in the majority of cases admitting a sound one-fifth of an inch in diameter—though, in a large minority, one from one-fifth to one-sixth of an inch only can be easily passed. I therefore regard a diameter of one-fifth of an inch as ample for the removal of stenotic sterility and dysmenorrhœa. I find the external os admits a dilator one-fifth of an inch in diameter and upwards—in some cases as high as one-fourth, or even three-tenths, of an inch—but, as a rule, I think one-fourth of an inch sufficient for the purpose. It is, of course, to be understood that no nar-

---

* The circle is smaller than the ellipse, in the proportion of 144 to 147.
† Circle to ellipse as 72 to 75.

rowing of the canal exists between the two ora. Since, however, there may be some degree of stenosis for the menstrual fluid, while not for the sound, it is sometimes judicious (and especially if congestion of the cervical lining membrane co-exists) to increase the dimensions just named, by the use of a dilator of the next larger size. I do not assert that the preceding dimensions are always required in the treatment of stenotic sterility and dysmenorrhœa, for they are not; nor that they are never to be exceeded; but that in almost all cases they will be found sufficient.

Should this precise specification of dimensions seem too minute for practical purposes, we must remember that dimension cannot here have a less important relation to function than elsewhere; and that enlarging the internal os to the diameter of half an inch, as is often done by the deep incision, is, as has been seen, like permanently dilating the urethra (if it could be done) to the size of the small intestine. And the importance of making an incision of the internal os, with a precise intention, and a precise knowledge of the mode of accomplishing what is intended, may be understood, when I state that if the circle representing its area in the imparous woman be increased equivalently to surrounding it by a ring only one-thirty-fifth of an inch wide, its area is increased as forty-nine to twenty-five, or almost exactly doubled. Or if an incision be made on each side of it to the extent of half of a line (one-twenty-fourth of an inch), and it then be dilated to a circle, it is increased two and a half times. And if the cut should extend one line to the right and the left, or the added ring were one-twelfth of an inch wide, the area would be increased more than four times and a half. This last increase is far more, in my experience, than is ever required in stenotic sterility and dysmenorrhœa. Peaslee now considers

SUPERFICIAL TRACHELOTOMY,

his own operation:

"Desiring to restrict the operation of trachelotomy in the treatment of stenotic sterility and dysmenorrhœa within the limits actually required, I, some ten years ago, devised, and brought before the New York Obstetrical Society* a series of five steel cervical dilators, to be used instead of incision, where the stenosis is slight and the cervix is normally soft and pliable. These, in shape and size, have a precise reference to the dimensions of the cervical canal, and especially of the two ora uteri, as already specified; and each is guarded by a bulb, so as to project through the internal os into the uterine cavity only about one quarter of an inch.

But finding that almost all cases of stenosis of the cervical canal are relieved more promptly, more permanently, and also with less pain, by incision, or this together with dilatation, than by any form of dilatation alone, I next endeavored to restrict the extent of the incision within the absolutely necessary limits, having determined them approximately by the preceding facts and calculations. To this end I devised a new method, and an instrument for executing it, which I also laid before the New York Obstetrical Society about eight years since; but the former was so simple, bloodless, and unpretending, in comparison with the procedures of Simpson and Sims, that it excited but little interest. Meantime, however, it has been sufficiently tested, I think, by myself and my pupils in different parts of the country, to entitle it to a more general notice.

Since the superficial incision, as suggested by myself, has for its direct object merely the removal of stenosis of the cervical canal, and is therefore proposed for the treatment of stenotic dysmenorrhœa and sterility only, it is previously to be decided whether stenosis actually exists. And the following propositions will aid in settling this question, it being understood that the exploration is to be made at least four days after, and at least three days before, the catamenial flow.

A. *Respecting Stenosis of the Internal Os.*

1. If a sound one-fifth of an inch in diameter passes easily through the cervical canal, there is no stenosis at the internal os, and no incision is there required. This is the size, therefore, of my large sound.

---

* Also described in the N. Y. Medical Journal, July, 1870, p. 478.

2. If a sound one-sixth of an inch in diameter be easily passed, as above, there is no absolute, though there may be relative stenosis of the internal os; *i.e.*, there may be stenosis for the passage of a fluid, though not of the sound; and an incision to one-fifth of an inch may be required, but not unless the symptoms indicate it.

3. If the sound easily passed be but one-seventh of an inch in diameter, and there are no symptoms of stenosis, no incision of the internal os is required. This is the normal size in the imparous woman, and the average size of Simpson's sound.

4. If a sound but one-eighth of an inch in diameter cannot be passed through the internal os, there is either stenosis, or, what is very much more probable, one of the flexions. Prove, therefore, that there is no flexion in this and every case in which a sound of any size does not traverse the internal os, before operating for stenosis. I consider an internal os of one-eighth of an inch, or less, to be stenotic. Chrobak's highest limit for stenosis of the internal os is one-tenth of an inch (two and a half millimètres).

B. *Respecting Stenosis of the External Os.*

5. On the other hand, there is no stenosis of the *external os*, if a sound one-fifth of an inch in diameter easily traverses it. If there be congestion of the lining membrane, however, there may be stenosis, practically, in respect to conception; and the operation somewhat enlarging it (to one-fourth of an inch or more) may be required.

6. If the external os will not easily admit a sound one-sixth of an inch in diameter, there is probably stenosis in respect to conception, and the operation is required. If not more than one-seventh of an inch, the operation will also probably be required for dysmenorrhœa.

7. In case of operation, the whole cervical canal must be made still to retain the normal fusiform shape as far as possible.

I. My *method* consists in incising the internal os, if the stenosis exists at that part—and the external, if at the latter—to such an extent as to give to both their precise average dimensions in the parous woman—neither more nor less—and, of course, also overcoming any other point of stenosis existing anywhere else in the cervical canal. In cases complicated with congestion, however, I have shown that a slightly larger opening may be required; and, therefore, that the limits may extend beyond one-fifth of an inch (to nearly one-fourth of an inch) in the case of the internal os, and to three-tenths of an inch, and possibly more, of the external.

I do not, therefore, incise the internal or the external os to a given depth in all cases; but, taking them as I find them, cut just enough to give them their average normal size in the parous uterus. This is seldom one-half a line and often not more than one-third of a line for the internal os, or more than a line for the external. But, of course, there is far more variation in the latter. If the internal os admits a sound of but one-eighth of an inch in diameter, a cut on each side of nearly half a line (but three-eightieths of an inch) is required; and if but one-tenth of an inch in diameter, it must be one-twentieth of an inch deep on each side. The incisions are of precisely the same depth on the two sides.

Since the lining membrane at the internal os is at most one-twenty-fifth of an inch thick, it is seen that I generally do not cut nearly through it. Indeed, when the os is but one-eighth of an inch wide, I cut almost through the membrane; and when one-tenth of an inch, I divide it, and one-hundreth of an inch into the tissue beneath it.*

II. The *instrument* devised to secure this effect consists of a flattened tube, containing a blade. The former is eight inches long, and seven-sixteenths of an inch wide, except its terminal one and three-fourths of an inch, which has a width of but one-eighth of an inch, as shown in Fig. 7. This portion is made curved by some instrument makers, which is not an improvement. The blade is of such a width as to slide accurately within the tube, having a nut and a screw attached to its proximate extremity to guage the extent of its passage into the

---

* The details of all the preceding calculations are properly omitted here; as a slight acquaintance with mathematics will enable the reader to verify them.

cervical canal, and a blunt point, and lateral cutting edges for one and five-eighth inches at the distal end. There are two blades for each instrument, the cutting portion of one being

Fig. 7.

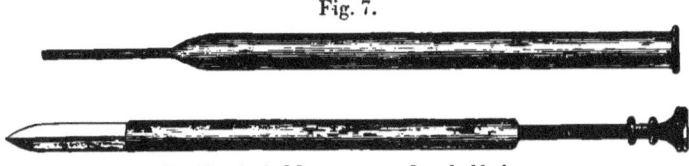

Dr. Peaslee's Metrotome.   One-half size.

one-fourth of an inch wide, and of the other three-sixteenths of an inch. If the stenosis is confined to the internal os, the narrower blade alone is used. If both ora are contracted, the wider instrument is passed through the external, and the other blade then introduced, and the inner os incised by it; and, in cases of decided congestion, the wider blade alone is sometimes used for both ora. In this case, a sound one-fifth of an inch in diameter is easily passed through the inner os; while, if the smaller blade had been used, a considerable force would be required to carry it through.

In hospital practice, I place the patient upon the side, use the duck-bill speculum, hold the cervix by means of a uterine tenaculum, pass the tube into the canal up to the shoulder, and, therefore, one-quarter of an inch *into* the uterine cavity through the internal os; when the blade, previously guaged, is introduced into the tube, and carried up the cervical canal as far as is required to overcome the stenosis. My large sound (No. 10 American scale), or, still better, the conical dilator of the proper size, is then passed up the canal, and the operation is completed. In private practice, I generally place the patient on the back, and pass the tube into the cervical canal precisely as I would Simpson's sound; and then pass the blade through it, as just described.

If the external os is too narrow for the admission of the extremity of my instrument, it may be enlarged by the introduction—generally one-eighth to one-quarter inch is far enough—of a narrow-pointed bistoury. I have not found the internal os too narrow to receive it, except in cases of flexion, or of previous traumatic injury of the cervix.

The changes in the whole uterine cavity from this operation, are shown by Fig. 9. Respecting its dangers I have but little to communicate. The hemorrhage following it seldom exceeds one or two drachms, and never requires any special attention. The pain is very slight and merely momentary, and no anæsthetic is ever required. The medullary structure of the cervix never being cut into, pelvic cellulitis and peritonitis do not ensue. The only exceptions to this statement, in nearly 300 cases, are one case in private practice, in which some febrile reaction and uterine tenderness ensued, which subsided entirely, without cellulitis, in four days; and two cases in the Woman's Hospital of slight cellulitis. But both the latter were patients who were known to have had cellulitis a short time previously; and I was obliged by some peculiar circumstances to operate sooner than I otherwise would have done. The final results were precisely as desired in each of these three cases. Otherwise I have never had any unpleasant symptoms follow the operation; and the only precautions taken are to keep the patient two days, and sometimes three days, in bed; and not allow her to walk out under a week. I used the dilator every second day after the operation for a week, and two or three times more once a week. I have very often performed the operation at my office on residents of the city, and sent the patients home to bed after half an hour's rest, and have never had to regret it. (I decline to operate within four days after or six days before the catamenial period.)

I claim for the method just described the following recommendations in the treatment of stenotic sterility and dysmenorrhœa:

1. It aims to restore the normal dimensions, as existing in the parous woman, throughout the cervical canal—nothing more and nothing less—unless when a slight exaggeration of size is required on account of co-existing congestion.

2. It effects this object definitely and with certainty, and with incisions exactly symmetrical, or equal on the two sides.

3. It gives no danger from hemorrhage, since the arteries nearest the internal os, if that is to be divided, are never reached, and the whole thickness of the lining membrane

even is generally not divided; and there are no arteries within the portion divided at the external os.

4. There is no danger of pelvic cellulitis, except in those patients in whom the least operative interference with the cervix, or the use of the sound, or of a sponge-tent, will produce it. I consider the operation less dangerous in this respect than the last-mentioned.

5. There is no danger of septic peritonitis, since the medullary substance is not reached by the incision.

6. It does not produce sterility, or tendency to abortion, by mutilating the cervical canal. The changes it produces in the latter, as compared with those from the operations of Simpson and Sims, are shown by Figs. 8, 9, 10 and 11.

7. It removes stenosis perfectly, and in most cases permanently, since there is very little tendency to closure of the slight incision made. I have had to repeat the operation only twice in my practice, except in cases in which there is cicatricial tissue to be divided; as after imperfect and partial closure, following rupture of the cervix in parturition, or ensuing after Simpson's or Sims's operation. Here the operation will usually have to be repeated in a year or two, unless pregnancy should

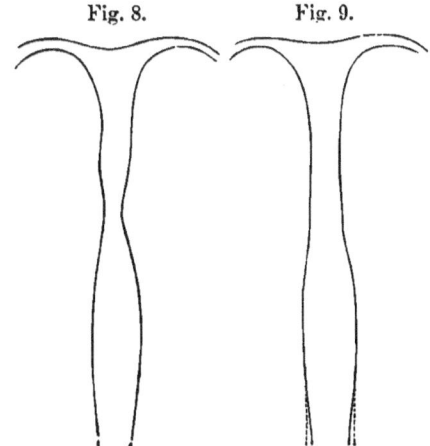

Fig. 8.   Fig. 9.

Normal uterine cavity.   Do., modified by Peaslee's method.

occur; an event not to be expected in such cases, as we have seen.

Finally then, since my experience has shown that a diameter of one-fifth of an inch for the internal os, and one-quarter to three-tenths of an inch for the external os is sufficient in the treatment of stenotic sterility and dysmenorrhœa, I suggest the disuse of Simpson's and Sims's operation in the treatment of

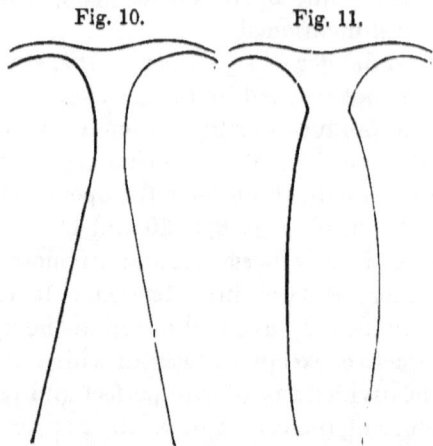

Fig. 10.   Fig. 11.

Uterine cavity after Sims's operation.   Do., after Simpson's operation.

these conditions; and the substitution of a milder, safer, and more efficacious method—of which, perhaps, my own is, however, merely the forerunner. At least, further experience in the line I have indicated will doubtless afford still more accurate conclusions.

There is another instrument, invented by Dr. Octavius White, of New York, nearly fifteen years ago.

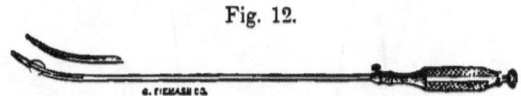

Fig. 12.

This instrument is the safest ever invented, and incises the cervix and inner os as much as is ever necessary, namely, one-fourth of an inch. The instrument, when the blades are withdrawn, looks like a long uterine sound. It has two small semicircular or crescentic blades concealed near its point. It is used with or without a speculum, and in any position the woman may assume. It is introduced up through the os internum, the two blades are thrown out by an action governed by a screw in the handle, and it is then withdrawn, cutting an opening only *one-fourth* of an inch, the normal size of the canal. I prefer to use this instrument through my speculum, for while it is nearly as convenient as Sims's, it needs no assistant. This is a bivalve speculum, the upper blade shorter than the lower—a very great improvement, as the practical operator will soon ascertain. I have lately improved upon it by changing the shape of the blades and inserting a small hook near the opening of the upper blade, for the purpose of retaining a tenaculum *in situ*. The following cut shows my speculum with the tenaculum.

Fig. 13.

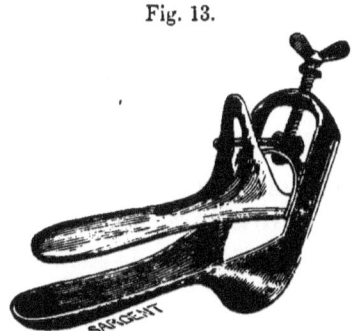

Hale's Expanding Speculum.

After the operation of incision of the cervix it is necessary to introduce some kind of a tent or plug to prevent contraction during the healing process. Sponge tents or laminaria are not always safe. I prefer the slippery elm tent, carbolized made of the proper size, as mentioned above. This should be removed in twelve or twenty-four hours and a plug of firmer material substituted. I have used Chambers's intra-uterine stem pessary in some cases. They are effectual if tolerated.

Fig. 14.

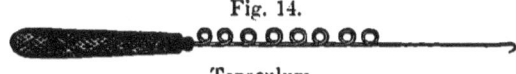

Tenaculum.

Dr. Sims figures an instrument which has some advantages

over Chamber's, in that it is tubular and self-retaining. It allows the secretions from the cavity of the uterus to pass through it. It is about two inches long; it is introduced with the wings drawn into a straight line by means of a stilet, as shown in the figure. As soon as it is passed to the requisite depth the stilet is withdrawn; the wings spring back into the cavity of the uterus; the os internum grasps the instrument at its bifurcation, and the lower end rests outside of and against the os tincæ. It should be made of vulcanite. It should be worn, if possible, through one menstrual period.

Fig. 15.

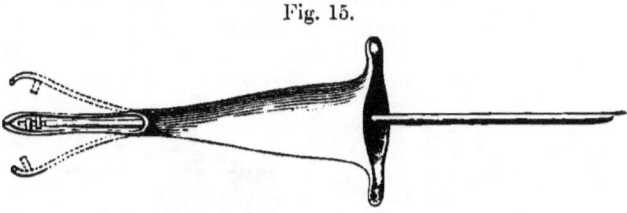

Dr. Coghlan's plan of using a tube of sheet-lead answers a very good purpose. Any physician of ordinary mechanical tact can make one in a few moments, and thus possess at hand a simple and inexpensive instrument.

Many gynæcologists prefer, instead of dilation by tents, or cutting instruments, the forcible and rapid dilatation of the cervix. One method is that recently adopted by Dr. John Ball, of Brooklyn, N. Y., by means of bougies and dilators. He thus describes his method of operating:

"First evacuate the bowels very thoroughly beforehand, so as to prevent all effort in that direction for two or three days. I then place the patient upon her back, with her hips near the edge of the bed, and when she is profoundly under the influence of an anæsthetic, I commence by introducing a three bladed self-retaining speculum,* which brings in view the os uteri, which I seize with a double hooked tenaculum, and draw down towards the vulva, when I introduce a metal bougie as large as the canal will admit, followed in rapid suc-

---

* My improved expanding bivalve speculum is better.—*Hale.*

cession with others of larger size until I reach No. 7, which represents the size of my dilator. I then introduce the dilator and *stretch the cervix in every direction*, until it is enlarged sufficiently to admit a No. 20 bougie, which is all that is generally necessary. Then I introduce a gum-elastic uterine pessary of about that size, and retain it in position by a stem, secured outside of the vulva, for about a week, in which time it has done its work and is ready to be removed."

Dr. Ball says, "The operation is not only applicable to all cases of constriction of the cervix uteri, but its crowning glory consists in the complete and radical cure of *flexions*, for which there had previously been no really satisfactory treatment." . . . "In cases of flexion the relief is obtained by the straightening of the canal, which is produced by a change of the muscular tissues of the cervix from an abnormal to a normal condition. In the rapid dilatation of the parts, the constricting fibres are of course lacerated to some extent, and in healing up around the pessary must necessarily conform to their new relation."

Dr. Ball in his pamphlet gives several remarkable cases of cure, illustrative of the value of this method. He says he has never seen any dangerous symptoms result from this operation. I certainly would advise a trial of this method, in preference to Sims's method of incision of the cervix. He also enthusiastically recommends this operation in *endocervicitis* and *conical cervix*.

A better instrument for dilating a constricted or flexed cervix is the one invented by Dr. A. S. Hunter, and which he calls a "Flexion Straightener and Dilator."

This instrument—the mechanism of which will be readily understood by referring to the cut—offers advantages when used either to overcome uterine flexions or to effect the dilatation of the uterine canal, which may be stated as follows:

1. Its peculiar point, because of its short curve, is compelled to traverse the convex instead of the concave wall of the flexion, and therefore, in the lesser flexions, it does not impinge against and become obstructed by the concavity of the flexed walls, as do the points of the other instruments thus far devised

for this purpose. To overcome the more pronounced flexions, this curve serves a like and even more useful purpose. In this latter class of cases the instrument may be used in the following manner:

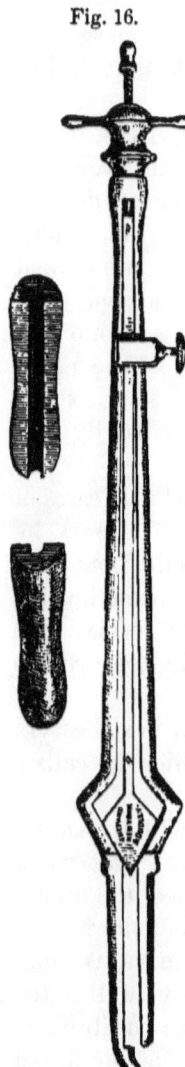

Fig. 16.

Hunter's dilator.

Having placed the patient under the influence of an anæsthetic, draw down and secure the uterus with a tenaculum. Now pass the point of the instrument up to the obstruction, with its longer finger opposed to the concavity of the flexion, slightly separate its fingers, and after allowing them to so remain for a short time, bring them together; the point will now pass still further on in the canal, and by patiently repeating this process, the object sought may generally be attained.

An easier, more conservative, and therefore better way to accomplish the same end in this latter class of cases, is, to first dilate the flexed canal without straightening it. This may be accomplished by the use of flexed tents of laminaria of small size, which have been soaked in an aqueous solution of carbolic acid, and which have been compelled to maintain, while drying, the required curve. One, two, or more of these may be introduced without difficulty, for there is less resistance to the introduction of a tent in sections than as a whole. Other sections may be added to those already inserted after they have become dilated; the patient being in the meantime in bed and under the influence of opium.

Again, the curved point of this instrument not only serves to facilitate its introduction in cases of flexions and cervical constrictions, but when inserted this same point opposes in the most satisfactory manner the concavity of the canal, and prevents the injury which might otherwise result to the organ when the fingers of the in-instrument are forcibly separated; further, while the presence of this projecting point within the uterus can under no circumstances prove objectionable, it may serve still another purpose in preventing dilatation too near the fundus.

2. The straight fingers of the instrument, together with their straight handles, render it capable of effecting dilatation in the antero-posterior direction. Does it not follow, therefore, that it is capable of straightening flexions?

3. As dilator it is also easily introduced, from the fact that its fingers are but 5-32ds of an inch in diameter, and also from the fact that its point follows and closely hugs the convex, and therefore does not impinge against and become obstructed by the concave wall of the canal. The fulcrum and lever speculum, if the uterus be either high up, flexed or anteverted, will also facilitate the introduction of this instrument.

4. Its capacity, notwithstanding its slender fingers, to effect dilatation, is secured by the close proximity of the power to the resistance.

5. Its elastic handles secure for it an elastic pressure with which to overcome flexions or cervical constrictions, or if it be desired to employ an unyielding force, this can be done by the use of the screw at the side of the handle. The same screw may be used to so change the direction of the fingers of the instrument that they will not only dilate the external and internal os equally, but either, more than the other, at pleasure; again, the elasticity of its handles enables us to estimate the force that is being exerted.

Finally. The straight fingers of the instrument, together with their straight handles, enable us to effect dilatation in any direction, therefore the points of bearing can be shifted so that the prolonged pressure on any given point may be avoided, and after the dilatation of the canal has progressed sufficiently to admit the gloves, which are then to be added to the fingers of the instrument, the pressure will be still further distributed, as well as its capacity to effect dilation increased. The instrument may therefore be made available to dilate the female urethra or the os uteri, to facilitate the removal of polypi or portions of retained placenta after miscarriage.

This instrument may be obtained of Shepard & Dudley, or Boericke & Tafel, of New York.

DISPLACEMENTS OF THE UTERUS.—Any of the displacements of the uterus may be a cause of sterility. In the various *flexions* of that organ, the bent condition of the cervix causes a stricture at the point of flexion. This stricture may form an obstruction to the entrance of the spermatozoa. (See Atresia, Occlusion.)

In *versions* of the uterus the os uteri is often placed in such situations that the seminal fluid cannot readily enter into the cervical canal.

ANTEFLEXION AND ANTEVERSION.—"Anteversion or anteflexion of the uterus may be a cause of sterility, but not more so than uterine diseases in general. Of thirty-three patients affected with this displacement, thirty were married, of whom *three were sterile*. The remaining twenty-seven had given birth to seventy-three children, and had had twenty-one miscarriages; or, whose labors amounted to 2.7, their miscarriages to 4.7 to a marriage; while one in 8.5 of the total number proved sterile" (*West**). "When *anteflexion* is so considerable as to nearly obliterate the uterine cavity at the point of flexion, conception is rendered very improbable; but still it may take place if no structural occlusion or atresia has recurred at the above-mentioned point, for there may be space enough left to allow passage to the spermatic fluid, and unless the uterus be held by false membranes, the flexion may be lessened in a horizontal position of the pelvis during coition by the weight of the fundus, and thus the occlusion of the cavity be diminished, and the impregnating fluid be allowed to pass. Considering, therefore, the frequency of sterility in the females affected with anteflexion (Mayer found 60 anteflexions in 272 sterile females), other circumstances should be taken into consideration, especially derangements of menstruation in the form of dysmenorrhœa or amenorrhœa; perhaps, also, the altered

---

* Diseases of Women, p. 168.

position of the vaginal portion (the os uteri looking backward) as well as a certain amount of cervical catarrh." (Kolb.)

Treatment.—The surgical treatment of *anteversion* is not attended with much success. The use of pessaries generally accomplishes but little in this displacement; but in exceptional cases they are of benefit, and should be tried. They act, not by rectifying the displacement, but by simply lifting up the uterus, and diminishing pressure on the bladder. *Anteflexion* is more apt to cause *sterility* than *version*. In rare cases, if we have in view only the cure of sterility, Sims's operation may be tried, as described by Thomas, which consists in *cutting through the posterior wall of the cervix*, thus converting a crooked into a straight canal, and affording a passageway for the seminal fluid. If this is objected to (as it is now very generally), the use of the intra-uterine pessary of Peaslee, Simpson, or Detschy may be tried. If these can be tolerated, the canal may be straightened. The use of intra-uterine stem pessaries is nearly abandoned by gynæcologists. The recent discussion in the American Gynæcological Congress showed that to be the case. I have, however, benefited several cases of anteversion and flexion, by the introduction of *Chambers's* intra-uterine *split*-stem pessary, which is the most efficient instrument yet invented, and the least liable to cause irritation.

Fig. 17.

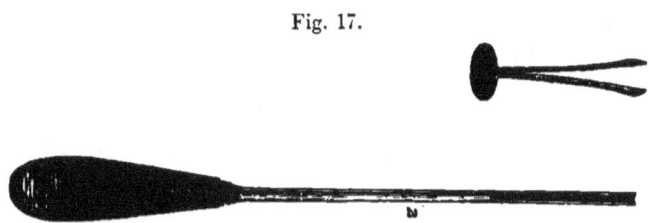

Since the first edition of this work appeared, I have used with perfect success, in the treatment of several old cases of retroflexion and anteflexion, a stem pessary invented by E. H. Sargent, of Chicago. It was suggested by Dr. T. G. Thomas's intra-uterine galvanic stem pessary, which consists of a button and a stem composed of alternate balls of zinc and copper, on a wire about two inches long. While this instrument may do very well for the purpose of stimulating the diseased membrane in cases of mucous cervicitis, *it sends no current through the uterine walls*. In order to effect this, Mr. Sargent invented an instrument whose split-stem was composed of copper, and the "button" of a ring of zinc, as figured below.

When this is introduced, the galvanic current passes in a circuit from the end of the stem through the uterine walls to

Fig. 18.

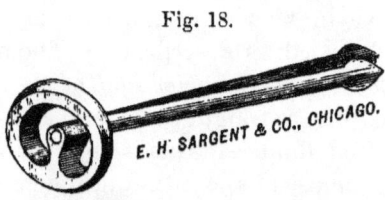

Sargent's Galvanic Stem Pessary.

the zinc ring. That this is not merely theoretical, my own experience has demonstrated, for I have cured in two months cases of retroversion and anteflexion, that had resisted Chambers's and all other kinds of pessaries, in the hands of the most noted gynæcologists of New York, Cincinnati, and Chicago.

Better often, than any intra-uterine pessary, however, is the simple bougie. or tent, made of dry *slippery elm bark*, such as is found in the shops. They are easily made with a penknife, and smoothed by means of a piece of glass or fine sand paper. They are very easily introduced after they have been soaked in water a few minutes; a mucilaginous substance oozes out of them, and in this condition they will slip into the flexed uterus without difficulty. I usually make them about two inches long, and of such size as seems best suited to the cervical canal. The point should be tapering, the outer end blunt, with a strong thread attached. This end can also be fixed in a ball or button, so as not to irritate the vagina, but it usually suffices to place a cotton ball (with a string attached) under the end. These bougies are to be placed *in situ* every few days, and removed in twelve or twenty hours, or sooner *if they cause pain*. The thread attached allows them to be removed by the patient. It requires two or three months to cure an ordinary case of *anteflexion* by these tents. Dr. Byford, of Chicago, speaking of the bougies, says: "The *modus operandi* of this instrument is different from the sponge, or seatangle tent. It dilates the cervical canal a little, but no forcing. It produces an influence upon the nervous vascular condition of the mucous membrane of the cervix which assists in removing the pathological state."

Fig. 19.

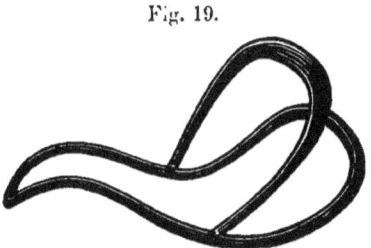

Thomas's Anteversion Pessary.—(Old.)

I have succeeded in curing several cases of sterility due to *anteflexion* by the use of Dr. Thomas's anteversion pessary. It lifts the fundus from off the bladder, and allows the cervix to assume its proper position. If the cervix has not acquired a permanently bent condition, a cure of the sterility will result.

Fig. 20.

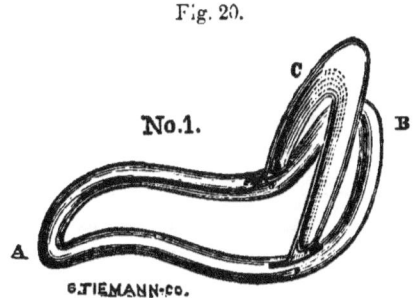

Thomas's New Anteversion Pessary.—(Shut.)

Introduce the pessary shut as in No. 1, then pull down the loop C until it slips under the cervix, then push it up between the bladder and the uterus. The end B, will lie behind the cervix, the loop C in front, and the lower end A will project

Fig. 21.

Thomas's New Anteversion Pessary.—(Open.)

under the arch of the pubes. (See Dr. Beebe's paper and treatment further on).

RETROFLEXION AND VERSION OF THE UTERUS may cause sterility.

According to Kolb, a retroflexed uterus may frequently conceive. Still Meyer records 36 cases of retroflexion in 272 of sterility.

"Retroversion," says Dr. Simpson, "is a cause of sterility, as shown by impregnation taking place after the displacement is rectified."

With reference to the influence of retroversion or flexion on fecundity, West says: "Of twenty-three affected with this displacement, four gave birth to live children after the womb had been misplaced, one had five children in spite of a retroversion of fifteen years. In one of the above four, pregnancy was preceded by the replacement of the organ; but in the other two, not only was the womb misplaced at the time of conception, but was ascertained to continue so after delivery. Four having given birth to living children, miscarried after the development of symptoms of uterine displacement, and in one of the number miscarriage had twice occurred, while fourteen, having previously given birth to one or more living children, had passed more than a year since the commencement of the symptoms without conceiving. In three of this number, however, though still within the childbearing age, conception had not taken place for two years in one instance, and for four years in the other two, previous to the commencement of the symptoms of the misplacement of the womb."

Diagnosis of Uterine Flexion and Version.

In most cases it is not difficult to diagnose a flexed or verted uterus by means of the finger or the sound. But there are cases where this means may fail us. For example, the fundus of the uterus may be retroverted and the cervix flexed—so that the whole organ is curved into a crescentic form,—the os pointing *backward*, as well as the fundus. The same may occur in anteflexion and version. In such cases a diagnosis may be made by means of a recently invented instrument manufactured by Dr. Jennison and called an Exploring and Indicating Sound, which contains valuable and remarkable qualities never before embodied in any for similar uses. In explorations of the uterine canal and in the diagnosis of malformations, growths, displacements, and, to a certain extent, as a repositor, there seems abundant reason for the belief that it is possessed of peculiar and positive value.

In its construction a number of light steel springs about fifteen inches in length are arranged upon and parallel to each other, united at their ends and placed within a small metal tube, which surrounds them, with the exception of about three inches at each end. One end of this tube is covered with hard rubber of size and form to constitute a convenient handle which allows the instrument to rotate easily within it, affording complete freedom of movement while being introduced, or it may be held above or below the handle if freedom is undesirable. The ends are each of about the diameter of Simpson's Sound.

The whole of the instrument except the handle being covered with a delicate flexible rubber sheath, is protected from the intrusion of fluids, and in all respects complete and convenient.

Its construction being understood it will be evident that any simple or single curve made in either of the flexible ends will be reproduced in an inverted form at the other; that an S, or double curve in one end will cause the other end to become straight, and that the instrument, while able to conform its distal extremity to the uterine canal, whether normal or abnormal, will reveal its real form at the proximal extremity.

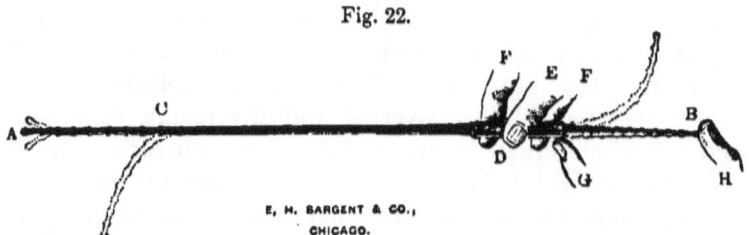

Fig. 22.

Fig. 22 is a representation, the dotted lines showing some of the almost unlimited number of positions of the ends attainable by manipulation. In the diagnosis of displacements by the use of flexible silver instruments, their form, when withdrawn from the os, indicates little or nothing because of straightening; not so, however, with this instrument, which at each movement of introduction or of withdrawal, indicates at the exposed end the form of the covered one.

In the use of any metal or partially flexible Sound in a canal whose axis does not correspond exactly to its own, the Sound overcomes resistance to its advancement by compelling the canal to assume its own shape; with the New Instrument, on the contrary, an undulatory movement or a slight increase of the curve already indicated is obtained by gently manipulating its proximal end, so that it may be made to pass where other instruments would be excluded.

SUGGESTIONS RELATIVE TO USING. (*See Fig.* 22.)

Hold the instrument firmly by the handle B in the right or left hand, as may be most convenient, the thumb E being uppermost, the fingers F F underneath; introduce the end A, and with the index finger and thumb of the other hand in the positions G H, it will be easy to manipulate the end B so as to obtain any required curve, combined with whatever of undulatory or worm-like movement may be useful while gently pressing the instrument forward.

Fig. 23.

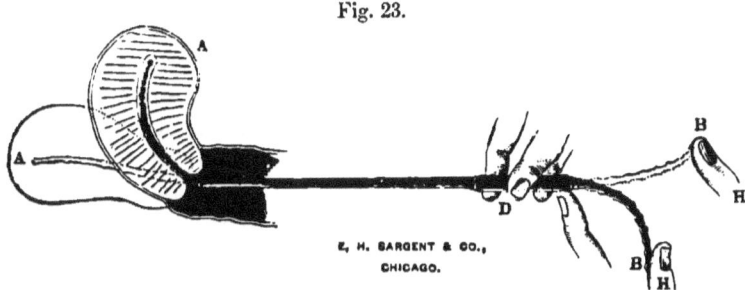

Fig. 23 shows the manner of finding a flexion of the uterus, and the dotted lines the method of replacement by reversing the curve of the end with the thumb, causing both ends to assume the new positions represented by the dotted lines, thus carrying the uterus to its normal position. Whatever the flexion or version hold the instrument firmly by the handle and with the other hand manipulate the end B in such a manner as will evidently be required from its position.

Instruments of sufficient delicacy for use as explorers do not possess strength for use as repositors in *all* cases. The instrument for general use will combine as far as possible adaptability to both purposes, while there will be a supply of such as may be required by specialists and others where greater delicacy or unusual strength may be needed.

Fig. 24.

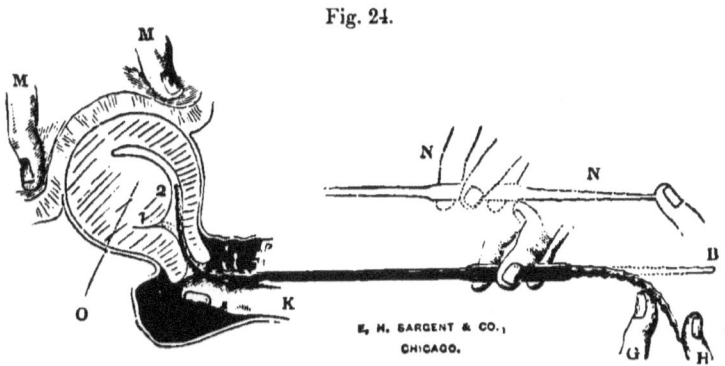

Fig. 24 illustrates its use in diagnosis of growths within the uterus, by manipulating the proximal end, as at G H, and the readiness with which it may be directed into and out of "pockets," as at 1, for the diagnosis of growths within the uterus.

Before introducing the Sound for this purpose, pass the forefinger of the left hand into the vagina in contact with the cervix as at K; if there is anteflexion of the uterus push it away from the pubes and with the right hand in the position M M, feel its surface for external growths; if none are apparent take the explorer by the handle and introduce it with the fingers of the left hand in such a position at the proximal end that any disposition to curve will both be readily perceived and may be favored in turn by corresponding manipulation as at G H. Continue to insert the instrument until it reaches the point indicated by the numeral 1, when its distal end will form a double curve resembling the letter S, while its proximal end will be straight. (*See dotted lines B*). The curve of the canal having been explored to this extent proceed further by slightly withdrawing the Sound, when, if the canal is somewhat similar to the representation it may be readily penetrated along the passage 2, the proximal end of the instrument assuming corresponding curves until having reached the end of the canal it will again be straight, showing the existence of another S curve of the canal (N N).

Thus having ascertained that the outer wall of the uterus is of normal form and that its canal is formed as indicated by the exploration, it seems evident that its distortion is caused by a growth, as at O.

The Instrument has been thoroughly tested in my practice and that of several gynecologists who heartily endorse it.

My own observation goes to show that retroversion and flexion is oftener a cause of *miscarriage*, than sterility. I have several patients in whom the uterus is permanently retroflexed; the structural changes are such that no permanent straightening of that organ can be effected, yet conception occurs readily; but in spite of all the precautions used to keep the womb from retroverting, miscarriage usually occurs between the second and third month.

UTERINE CAUSES.

Treatment.—I have had considerable experience in the treatment of this form of displacement, and the result is that the mechanical treatment of *retroflexion*, when we wish to cure *sterility* from that cause, should consist in:

(*a.*) The straightening of the uterus at its point of curvature by means of Hunter's instrument, or Chamber's or Sargent's galvanic stem pessary, or the slippery elm tent, allowing them to remain in the canal of the cervix only a few hours every day; or

(*b.*) The continuous use for several months of Dr. Albert Smith's pessary, Dr. T. G. Thomas's improved double lever

Fig. 25. Fig. 26.

Thomas's retroversion pessary.   Albert Smith's pessary.

pessary of hard rubber, or Dr. A. R. Jackson's soft rubber pessary of the same shape as Thomas's, but with the wire framework left open at top and bottom; or the following

COMBINATION PESSARY.

There has recently been introduced by Dr. Studley (*Amer. Jour. of Obstetrics*, vol. xii, p. 39), in an elaborate paper on Flexions and Versions of the Uterus, a new pessary, or a combination of the retroversion and stem pessaries. It consists of a rubber band stretched across a narrow Albert Smith Pessary, through the middle of which band is thrust a vulcanite stem, varying from the straight to any curve desirable, with a button fixed to one end. Down the stem is placed another perforated screw-button, adjusted to a screw-thread on the stem at its base, by which means the stem was firmly fixed to the band, with its curve in anterior, posterior, or side direction, as desirable. In the button end of the stem a hole is drilled to the depth of third of an inch. This is for the purpose of receiving a wire probe held in the hand to facilitate introduction. The accompanying cut shows the straight stem instrument.

Fig. 27.

Studley's straight stem pessary.

Dr. Studley gives his method of replacing the uterus and placing his pessary as follows:

"With the patient in the semiprone position, and by the aid of Sims's speculum, the instrument is easily introduced. The first step is, of course, to anteflex a retroflexed womb, to retroflex an anteflexed one, and lateroflex in the opposite direction a lateroflexed one.

My method of doing this is as follows: Cut off the end of a No. 6 or 8 flexible bougie to the length of about three inches; around and close up to its open extremity wrap a band of adhesive plaster, a third of an inch wide, sticking its two ends together and allowing them to project a third of an inch. Grasping this projection with the dressing forceps, and after dipping it in warm water, pass it into the uterine canal. It easily accommodates itself to the flexion. Now holding it *in situ* with the forceps, pass differently curved wire probes into the bougie, thus gradually straightening the flexion, and at last flexing it in the opposite direction by correspondingly turning the probes. I have these wire probes made smooth, nickle-plated, fixed to light handles, and armed with a little sliding vulcanite ball, to indicate where I have reached the full depth of the bougie. It is well to begin treatment with employing this flexing process alone on two or three different occasions before introducing the flexion stem pessary. Having got the womb flexed as we wish, the next step is to pass the wire probe into the hole of the stem, by which it can be turned in any direction on the india-rubber band, and introduce it as you would a curved sound into the uterine canal; at the same time holding the vaginal part of the pessary in such a way as to facilitate its introduction into the vagina. If we have a case of retroflexion, the vaginal part should be of good length and the band should be placed near the transverse bar, so as to carry the cervix backwards and upwards, and thereby accomplish an anteversion as well as an anteflexion. If we have an anteflexion to deal with, the vaginal part should be shorter and the band placed further down the arms, so as to admit of shortening the vagina and bringing the cervix forward. In a case of lateroflexion, the stem should be turned over the band, so as to fix it in the opposite direction.

The patient, in most cases, is allowed to be about and indulge in a moderate amount of exercise. As a rule, the pessary can be worn for a week or ten days without trouble. If pain or constitutional disturbance to any extent ensue, the patient is ordered to remove it at once—a thing she can easily do. On removal, an anteversion or retroversion pessary should be introduced. After a rest of a few days, the flexion pessary should be again introduced, to be removed as before, after a short period; and this mode of dealing should be continued in accordance with the requirements of each case. Such a treatment of a week's duration will accomplish more than six months of treatment by the sound, and be attended by far less annoyance and, I think, with far less danger. I have now treated a goodly number in this way, nearly all of whom have been greatly benefited, and the majority furnish striking evidence of its merits.

The principle of the action of the curved stem pessary is plain. When *in situ*, the atrophied wall of the womb is put upon the stretch, and the hypertrophied one is shortened and condensed by pressure. In the one, nutrition favored; in the other, absorption takes place. Many advantages are obtained from the elastic band. It keeps the stem securely *in situ* without rigidly fixing it, while at the same time it constantly, but gently, exerts an anteverting or retroverting force so desirable in connection with corresponding flexions."

As a large proportion of our school still protest against the use of pessaries, and allege that their use is mainly sanctioned by old school authorities, I here thought it best to quote the following excellent paper, written by one of the most clearheaded and thoughtful members of our school. It was published in the "American Homœopathist," and entitled "The Use of Pessaries," by Albert G. Beebe, M.D., Professor of Principles and Practice of Surgery and Clinical Surgery in Chicago Homœopathic College:

"It seems, at first thought, almost incredible that any great differences of opinion should exist among educated and experienced physicians regarding the value of pessaries in the treatment of uterine dislocations.

"That such differences of opinion do exist, even among well educated and skillful practitioners, we know; and when we reflect upon the amount of accurate *anatomical knowledge*, the clear *apprehension of mechanical principles*, the *inventive ingenuity*, and patient *adaption of means to individual cases*, necessary for the successful use of these various contrivances for maintaining the normal position of the uterus, we cease to wonder that so many fail utterly to comprehend the first principles of this branch of mechanical surgery. It may, however, be confidently asserted that no person has ever achieved marked success in the treatment of these conditions, without mastering, at least, most of the more important principles underlying the use of pessaries. In short these instruments are a *sine qua non* in the successful treatment of uterine deviations. I would not assert that displacements are never relieved or cured by other means, but that, *as a rule*, such mechanical appliances are necessary and indispensable. If our use of such apparatus is to be scientific and satisfactory it is necessary, *first*, that we have a thorough knowledge of the anatomy and physiology of the pelvic tissues, particularly, and their mutual relations to each other, mechanically and physically; *second*, it is necessary that we should carefully study the mechanism of the various displacements of the womb whether occurring as primary or secondary conditions, *i.e.*, whether the cause or effect of disease: *third*, we must consider what means we may avail ourselves of to correct these displacements the most naturally, certainly, and pleasantly.

"It is not the purpose of the present paper to dwell upon the 'firstly' and 'secondly' to any considerable extent but pass on to the 'thirdly' and attempt to give some estimate of the more valuable means for replacing and retaining in proper position a displaced uterus.

"We have, of course, two classes of pessaries, *i.e.*, those introduced entirely within the vagina and those extending outside the body and maintained by some external support. The former class are designated as 'vaginal pessaries' and the latter as 'stem pessaries.'

"It is not to be supposed, however, that the term 'vaginal' indicates that this class of pessaries depend upon the vagina for their support, but simply that they are contained entirely within the vagina. The tissues which support such pessaries are the bones and the muscular floor of the pelvis, while the vagina is allowed to remain as nearly in its normal position and is as little distended as possible. It is in the thorough understanding of these supporting parts and in the mechanical ingenuity to take advantage of their action that the secret of success in the use of such pessaries lies.

"Among the different forms of displacement I suppose we are bound, in defference to popular opinion, to consider first *prolapse*. Patients always, and physicians generally,—that is, so many of them as have only a general, vague knowledge of the subject—consider this as the chief if not the *only* form of displacement. What school boy has not heard of 'prolapsus uteri?' And what inventor and patentee of agricultural implements has not produced a combined abdominal supporter and uterine 'elevator' warranted to cure every imaginable form of 'female weakness?' As a matter of fact simple prolapse of the uterus is among the more uncommon forms of displacement. When it does occur in a pronounced form, especially if it shall have arrived at complete procidentia, it is an exceedingly troublesome condition to handle, and may defy not only all kinds of pessaries but all surgical expedients, whatsoever. This would be, it is true, an exceptional case, and would result from the displacement of other pelvic or abdominal viscera, or the entire loss of the more important uterine supports, as the perineum or the broad ligaments.

"Prolapsus is not unfrequently produced by relaxation or distention of the vesico-vaginal or of the recto-vaginal septum thus allowing their contents to force the bladder or the rectum into the vagina and even out of the vulva, dragging the uterus down at the same time. This prolapse of the vagina is, of course, greatly favored by rupture of the perineum, or by great relaxation of the sphincter vaginæ. In such cases the use of any form of vaginal pessary would be practically impossible for the reason that there is no support upon which it can rest.

"This would necessitate the resort to some form of stem pessary, and since it is somewhat problematical whether a uterus which has once been completely extruded from the body can be ever again depended upon to maintain its normal position unaided, the selection of a proper pessary is a matter of no slight importance. Even in much less aggravated cases the aid of such instruments will be required for a considerable time, and in order to be successful should be constructed upon strictly correct principles and accurately adapted to the individual so as to be worn with ease and comfort.

"All such instruments consist essentially of two parts. The uterine portion, or that which directly supports the womb, and the external portion by means of which the uterine portion is sustained and kept in place. In the first place it may be assumed as un-physiological that the uterus should be supported by any form of pressure brought to bear directly upon the os or lower portion of the cervix. Such pressure is always objectionable and injurious. Consequently all forms of pessaries which seek to support the womb by means of a cup designed to receive the cervix are radically wrong in principle and should be discarded. If we are asked why pressure by this method is harmful, the answer would be, because, as a clinical fact, we know that pressure applied to this portion of the womb does produce erosion of the os and painful inflammation even when a displaced uterus presses only against the vagina. Again, the uterus is naturally supported entirely by tissues attached above this portion; it is indeed suspended, instead of resting upon its apex, and the more nearly we can approach the natural method the greater will be our success.

"The use of a ring which shall encircle the cervix and so bring its pressure at the utero-vaginal junction is more philosophical, but is open to the objection that the uterine neck is liable to become strangulated in this ring, unless it be made so large as to unduly distend and weaken this portion of the vagina. Perhaps the least objectionable form would give us a stem having the curve of the vagina, forking so as to send one branch anterior to the cervix and a larger one posterior, both terminating in rounded crescentic masses whose concavities should not embrace more than two-thirds of the cervical circumference.

"As to the external means of support, we should seek some point or points of attachments where there will be as little liability to displacement or disturbance from the motions of the body as possible. One class of pessaries widely advertised and doubtless often sold, have a pad two or three inches in width to be applied over the hypogastrium and kept in place by a strap of elastic webbing passing around the body. To this pad is attached a curved rod or wire say eight to twelve inches long bearing upon its other extremity a cup or some such device for supporting the womb. It stands to reason that it would be practically impossible to make any such arrangement, resting upon such a soft and fluctuating foundation as the abdominal parietes, give any efficient support to the uterus with the disadvantage of such leverage against it. If the stem is supported upon rubber cords or tubes attached in front and behind to such a belt, it still is liable to be drawn backward or forward by any motion or displacement of this belt and thus derange the direction of the supporting force and consequently disturb the womb.

"If, however, two such rubber bands or tubes pass forward from the lower end of the stem, through the groins to meet two others, passing backward through the gluted folds, at the crest of the ilium, perpendicularly over the hip-joints, and there attached to the patients clothing or a simple band passing around the hips, we have reduced the tendency to displacement to its minimum.

"A pessary consisting of a stem, such as described above, and supported in this way, if of suitable size and form to be suited to the patient, might be, I can conceive, a very satisfactory instrument in such cases as cannot be managed by internal pessaries. The elasticity of the rubber supporting-cords being an important element in the apparatus.

"In some cases in which the perineum and sphincter vaginæ were not seriously impaired I have succeeded well by the use of a modification of the oval ring, making it somewhat rectangular at its lower extremity and at its upper half, double so as to send one bow in front of the cervix as well as one behind.

"The great majority of cases usually classed as 'prolapse' are essentially nothing more nor less than retroversion. The uterus is in such cases, it is true, somewhat lower in the pelvis than is normal, but the primary condition is the retroversion, and when this is remedied the prolapsus disappears. It matters little, however, whether we call the condition by the one name or the other since the treatment is about the same in either case.

"This brings us, then, to the consideration of the pessaries most useful for the treatment of ordinary cases of retroversion, or as some might prefer to call it the first stage of prolapse. The term retroversion is preferred by the writer because it is the precedent condition in nearly all cases, and in many, is unaccompanied by any marked descent.

"In the treatment of this condition, undoubtedly Hodge's closed lever, with its various modifications, ranks far above all others in general usefulness.

"Indeed, when properly formed and adapted to the patient, I should unhesitatingly pronounce it *the most perfect of all pessaries*—a model of scientific simplicity.

"In order to thoroughly understand the principle upon which this instrument acts, it is necessary to thoroughly comprehend the physiological relation of the parts involved. In front we have the pubic arch with the urethra emerging from behind the bone at the apex of this arch. Extending obliquely backward and downward and bounding laterally the vaginal orifice are the muscular columns of the sphincter vaginæ, terminating posteriorly in the perineum, and expanding laterally into the muscular floor of the pelvis. The anterior wall of the vagina is short and rises nearly perpendicularly to its attachment rather low down upon the uterine cervix, while the posterior vaginal wall, being much longer, passes back more horizontally from the vulvar commissure, then sweeps upward and forward to be attached rather high up on the cervix, nearly opposite the os internum. The ostium vaginæ thus looks downward and forward; the perineum approaches the pubis, making this opening much longer in its transverse diameter, the anterior and posterior vaginal walls being approximated in the same manner.

"A perfect pessary should correspond accurately to the natural curve of the posterior vaginal wall, and should lie within the vagina without distending it in any direction. Its anterior extremity, to the extent of half an inch or so, should curve downward, so as to engage slightly under the pubic arch, and should be just wide enough to be retained easily. It should also be so formed as to avoid producing pressure upon the urethra. When the pubic arch is narrow and the muscles firm, the Albert Smith modification works admirably; but in many cases this would not be retained, and it may even be necessary to make the front of the instrument quite rectangular and as wide as two inches, in extreme cases. The body, or horizontal portion, should be nearly straight, and this should merge into the greater curvature, comprising nearly one-half of the whole and forming something less than the quadrant of a circle. The superior extremity should be neither too narrow nor too pointed, but a rather flattened curve, transversely.

"When in position and subjected to pressure from above, the

anterior extremity is deflected downward by the pubic arch, and, as the body of the instrument rests upon the perineum or floor of the pelvis, the upper extremity is thus thrown upward and forward by a lever-action, and, of course, carries the uterus with it. The uterus, also being supported by its posterior vaginal attachment alone, and being held down somewhat by the anterior vaginal wall, naturally tends to drop forward by its own weight.

"Some care, and, perhaps, repeated trials, may be needed to adjust the length of the pessary, so as to accomplish just enough and not too much. In some cases the anterior vaginal wall and the round ligaments are so lax as to allow the uterus to rise up and ride over the top of almost any pessary that can be employed, however long. Indeed, if the pessary is so long as to straighten out the fornix, it prevents the pouching upward of the vagina behind the uterus, and defeats its own object.

"The Albert Smith pattern develops the lever-action much more fully than any other, and is, therefore, superior when it is practicable.

"Aside from the modifications of the Hodge's closed lever pessary, the one which would seem to most nearly fulfil the indications is the Cutter pessary as modified by Thomas. The objections to this instrument are that the pressure against the fourchette and the presence of the rubber tube in the cleft of nates, often cause excessive annoyance.

"However, as this would only be required when the vaginal pessaries could not be retained, it may fill a gap not otherwise so well occupied.

"The treatment of retroflexion will hardly require, ordinarily, pessaries essentially different from those employed in retroversion, but in flexions it will often occur that something besides mechanical supports will need to be employed before the uterus can be induced to maintain its normal rectitude. This remark applies especially to anteflexion which is, by far, the most exasperating form of displacement upon record,* and

---

* I am inclined to think I shall have to make an exception to this statement in favor of an occasional case of lateroflexion.

is also, unfortunately, the most common. At least such is my experience. It is predominantly a disease of the nulliparous uterus, though not exclusively so. It is generally a condition of mal-nutrition of this organ and very often associated with an obstinate form of anorexia and general mal-nutrition of the entire system.

"Such patients often succeed wonderfully in defying all efforts to make them eat or sleep; and are nervous beyond our most sanguine expectations.

"The uterus is excessively irritable and sensitive, as well as depreciated in vitality, flabby, and sluggish in responding to treatment; but when replaced in its proper position, exhibits a perseverence in returning to its distorted state that is truly astonishing and would be admirable were it exerted in a better cause.

"There are several reasons why the use of pessaries in this condition is especially difficult and unsatisfactory.

"In the first place it is difficult to construct a pessary which will pass up sufficiently far in front of the womb to hold it upright without unduly distending the vagina or being exceedingly awkward to introduce. In the second place the anterior vaginal pouch is often too shallow to allow a pessary to pass up in front of the uterus beyond the point of flexion, and in the third place, the womb not unfrequently acquires a fixed curvature which nothing short of direct extension by a ten pound weight would overcome.

"This naturally makes one think longingly of intrauterine pessaries, and if these uteri would only tolerate them, they would indeed be a God-send.

"So long as the patient will submit to their use, nothing can compare in efficacy to laminaria tents, dried in a curved form and introduced to expand and straighten. These may be followed by sponge tents, if necessary, and really constitute one form of intrauterine pessaries, when thus employed.

"It is not to be supposed, however, that anteflexions are not to be controlled by pessaries. They may be, in a large majority of cases, if we have the patience to find the proper instrument for each case. Before introducing a pessary, it is to be borne in mind that, without doubt, nine-tenths of the cause of flexions consists in undue pressure from above, produced, mainly, by tight clothing. This pressure must be removed as far as possible by loosening the clothing and, in some cases, by the aid of a proper abdominal supporter.

"In a few cases I have found a small oval ring, introduced in front of the cervix, to be retained and answer admirably. This is more likely to succeed if it is bent around the side of a cylinder, say an inch and a half in diameter, and introduced with this concavity forward. In most cases, this will get out of place; and to avoid this I have sometimes fixed a short projection upon the lower segment of the ring, (like the handle upon a hand mirror,) which being about half an inch in length, is held in the posterior commissure of the vulva and so prevented from displacement. Thomas, of New York, has evidently devoted a great deal of study and mechanical invention to the treatment of this form of displacements and has brought out quite a variety of pessaries for anteflexion and anteversion. Among these that which usually bears the name of Thomas's anteversion pessary is probably the most valuable. When the vagina is sufficiently capacious and the uterus heavy enough to bring it down into the grasp of the instrument this will often succeed admirably; perhaps more frequently than any other single form. It will often occur, however, that the flexed uterus will lie upon the top of the anterior bow of this pessary, or, indeed, of any other that can be devised.

"The latest form which I have devised, but which I have not yet had sufficient opportunity to test fully, is simply the modified Hodge's pessary, bent strongly upon itself, so that the upper, or uterine extremity, shall pass up in front of the cervix, instead of behind it. It seems to me to fulfil every indication; and, if a more extended trial shall justify my hopes and anticipations, it will prove to be 'the way I long have sought,' for reaching these cases, *tuto, cito et jocunde.*

"To enumerate and criticize the host of pessaries which have been brought forward by various persons, would require an expenditure of time and paper which the labor would not repay. Any one who has any practical familiarity with the disease and the principles of its treatment, will not be misled by such trumpery as many of these are.

"A successful pessary, let it be borne in mind, should be readily introduced and replaced; should not unduly distend the vagina, and should be light, cleanly, and not liable to become displaced or to irritate the parts. Some cases, which had defied other appliances, I have successfully treated by a stem pessary curving somewhat forward and terminating in a rounded, T-shaped head, and supported by four rubber elastics (as described when speaking of prolapsus); or, still better, having in place of the first six inches of the rubber tubes passing forward through the groins, firm elastic branches fixed to the lower end of the stem, thus securing the stem against being deflected backward out of its place, possibly behind the womb. This has the advantage that it can be readily removed and replaced by the patient each night and morning, thus securing cleanliness and avoidance of irritation.

"After all, when we have tried all kinds of pessaries; have dilated with tents; invigorated and stimulated the womb by local treatment, or the intrauterine electrode of the faradic or galvanic battery; have restored the general nutrition by food and internal medication, there are still some cases which will make us feel that much is yet to be desired in the way of means to satisfactorily handle all cases of anteflexion.

"One of the most important questions in the use of pessaries will naturally be: How can the practitioner accurately adapt pessaries to each patient, or change them from time to time, as the exigencies of the case may demand? This cannot be done so long as each instrument must be bought of the instrument maker at an expense ranging from half a dollar or a dollar upward, even as high as twenty-five dollars, and made by a mechanic who can have little conception of the principles of the apparatus, and much less of the case for which it is to be employed. A large proportion of the Hodge's pessaries, for example, are utterly worthless in the shape they are sold. The physician who would expect to *fit* a pessary to each of his patients requiring one, would need to have a peck basket full at his elbow; and, as no one but a specialist could afford to do this, the result is that very few pessaries are adapted to the patients who are condemned to endure them. The only method of solving this problem, in general, is to have pessaries made, or to make them, of some material which can be readily molded to any desired form.

"So far as I am aware, no other material is so generally available, so cheap and cleanly, and capable of being wrought into every possible kind of pessary, as gutta percha.

"The expense of the material is insignificant, and, with a little skill and experience in handling it, it can be made, with the occasional assistance of a little brass spring-wire, to answer for every kind of uterine dislocation. Hard rubber is also capable of being shaped within certain limits, if carefully heated, but cannot be lengthened or shortened, nor is it elastic enough to yield to the pressure of the organs, as gutta percha will do. It admits of a more perfect finish, and is entirely impervious to moisture, and is, therefore, even more cleanly than the other. A variety of pessary has been recently introduced, consisting of a rather stiff copper wire covered with pure soft rubber. These are flexible enough to be easily bent into any desired shape, and firm enough to hold that shape when introduced. If carefully guarded against becoming offensive from the secretions, they will certainly prove exceedingly convenient.

"This brings us naturally to consider the length of time a pessary should be allowed to remain *in situ*, without removal and examination. A great deal of abuse has been heaped upon pessaries in general, on account of the abominable abuses which have been perpetrated through their agency.

"No physician should leave a pessary in a patient's vagina without explaining clearly the necessity for its removal within a week or two at farthest, and *at once*, if it causes much disturbance.

"It is the duty of the physician to see to it that every pessary he introduces is removed as often as every week or ten days, and a day or two interval given before its reintroduction, especially if it produce any irritation; and any one who would permit a patient to wear such an instrument, no matter how simple, for months continuously, as has often been done, is guilty of the grossest malpractice.

"When it is possible to construct instruments so as to allow the patient to remove and introduce them whenever it may seem necessary, it is very desirable to do so, on several accounts. It may be necessary or desirable, sometimes, to erect the pessary as a barrier between husband and wife, but it would not be prudent in all cases to make it so too continuously.

"In a majority of cases where such mechanical aids are required for the correction of uterine deviations, they will, if skillfully managed, discharge their duty and may be discarded in the course of a few weeks, or a very few months at most. There may be cases in which pessaries may be required for a much longer time, or even for the remainder of life, but they are the exception and not the rule.

"Of course it is almost unnecessary to say that it often occurs that conditions of acute inflammation or excessive sensitiveness of the uterus will render the use of any form of pessary impracticable for the time being. These conditions must be overcome by other kinds of treatment until the use of mechanical aids will be tolerable.

"It is to be hoped the time has nearly come when one portion of the profession will cease to denounce all pessaries as mischievous and useless; and the majority of the remainder will no longer throw some kind of pessaries at their cases of uterine dislocation in a random sort of way, being guided only by a vague notion that some kind of appliance, called a pessary, might be of benefit, and ought to be tried.

"Somewhat more than a year ago, in an article upon the use of pessaries, I referred briefly to those available for the treatment of these displacements, and referred to a new pessary, which I had devised while preparing that paper, as follows: 'The latest form which I have devised, but which I have not yet had sufficient opportunity to test fully, is simply the modified Hodge's pessary, bent strongly upon itself, so that the upper or uterine extremity shall pass up in front of the cervix, instead of behind it. It seems to me to fulfil every indication; and if a more extended trial shall justify my hopes and anticipations, it will prove to be the way I long have sought for reaching these cases, *tuto, cito et jucunde*.' The object of the present article, in again referring to this subject, is to say that the experience of the past year and more has been to *fully realize* all the hopes above expressed; to explain a little more fully the shape of the instrument and the mechanism of its operation, and to touch upon some other points in the treatment of this form of displacement, *i. e.*, more particularly, *anteflexion*.

"To any one who is practically familiar with the treatment of this disease in its various forms and with the instruments heretofore available, I am sure no arguments will be required to prove the desirability of improved means. It cannot be said that the number of pessaries produced for this purpose has been very small, but the number of those which are of any practical value is exceedingly limited. Considering then that a large majority of all displacements belong in this class, and especially those of the most obstinate character are most often in young or nulliparous women, it seems evident that any important addition to the *armamentarium* of uterine surgery is by no means an insignificant matter.

"Among the pessaries which have been introduced to the profession for anteversion or anteflexion, the most prominent are Thomas's anteversion, Cutter's, Hurd's, and the various intrauterine pessaries. As to the latter there are at present but few gynæcologists who have the courage to use them in any case, and even their most enthusiastic advocates have never claimed their applicability except for a very limited number of selected cases. In these cases, and even in a much larger number, I have no doubt that the judicious use of laminaria, slippery elm or sponge tents, may be made to succeed much better and with less risk to the patient. Indeed this means may be employed alternately with a proper vaginal pessary, and, in many cases, is indispensable to the successful treatment. This is especially true of cervical flexions and of all cases where the organ has acquired a *fixed* flexion. The process of expansion and subsequent contraction of the uterine tissue, tends more directly and positively than any other measure to restore the normal form and texture of the organ.

"At the same time, the direct pressure exerted upon the uterine mucous membrane contributes powerfully toward the removal of the catarrhal inflammation, which is almost invariably and inevitably associated with this displacement. If not more than two tents are used successively there is but little ground for apprehension, so long as no acute inflammation is present, and, of course, no one would think of employing them if such were the case. We may say, then, safely, that intrauterine pessaries are not to be considered as practically available for the *support* of a displaced uterus. As to Hurd's pessary, which seems to have been approved by Thomas, it is evident from even a superficial observation that it could be made operative in but a very small number of cases, and those of the simplest and most tractable character, viz., where the flexion

Fig. 28.

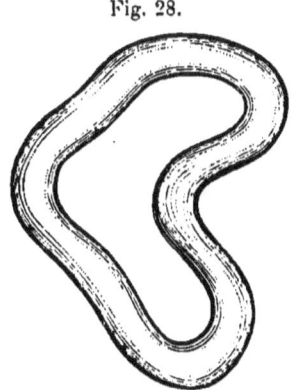

is corporeal, the vaginal portion long and slender, the uterus lies low in the pelvis and the vagina is capacious. Even in the most favorable cases the pessary is objectionable on account of distending the vagina and thus weakening its columnar character, and favoring the retention of the vaginal mucus from the extent of surface covered. These objections are more apparent from inspection of the instrument itself than from the appearance of the engravings in Thomas's work. It is difficult to believe that Cutter's pessary would ever be much employed for such cases. Except in a very few cases in which the perineum is very firm and approximates the pubis closely, it seems inevitable that sufficient tension upon the supporting band to produce any supporting action would only result in drawing the pessary downward and backward out of the vagina. Certainly the instrument would need to be made of a very different size and form from any I have seen in the instrument shops to allow of its being worn at all in this position, except in exceptional cases. Even were it entirely practicable and successful, the irritation and annoyance from the supporting band would make this instrument very objectionable to most patients.

"By far the best of its class has been that known as 'Thomas's anteversion,' consisting of an anterior uterine arch jointed to and supported by an Albert Smith pessary. In some cases where the uterus readily oscillates between anteflexion and retroflexion or anteversion it is certainly unexcelled by anything.

"It has done most excellent work, and is an honor to the accomplished gynæcologist whose name it bears.

"The disadvantages which must be admitted are, that it is often quite difficult to introduce and remove, and, therefore, not under the patient's control, it is only available in moderately capacious vaginæ and where the uterus lies rather low with a large anterior cul-de-sac, and even in these, it distends the vagina to an objectionable degree. In some cases the anterior arch is crowded back so as to seize the neck like forceps, producing severe pressure and pain. It will thus be seen that a large proportion of cases are excluded from the satisfactory use of even this pessary. The various designs which Thomas and others have brought out during the last few years, seem to be principally tentative, and only demonstrate the evident demand for better means for the treatment of these cases.

"The pessary which I herewith offer to the profession as meeting, as it seems to me, all the requirements of the case, is so exceedingly simple and unpretending that it seems almost unnecessary to occupy space in describing it. Considering the Hodge's pessary, as modified by Albert Smith, the most perfect instrument of its kind, I sought to devise one for anterior deviations, having this as a basis. It will be seen by reference to Fig. 29 that the lower, or perineal portion of the two are the same, and they hence have a similar lever action. To convert the *retro*-uterine into an *ante*-uterine supporter, it was only necessary to bend it forward upon itself and then curve its upper extremity slightly upward. Those I have used were made of the flexible wire covered with rubber tubing, or of hard rubber. The former are more convenient when frequent changes in shape are necessary; the latter where the pessary is to be worn rather constantly, because easily introduced and removed, more cleanly and less irritating. The advantages which this instrument seems to me to possess are these:

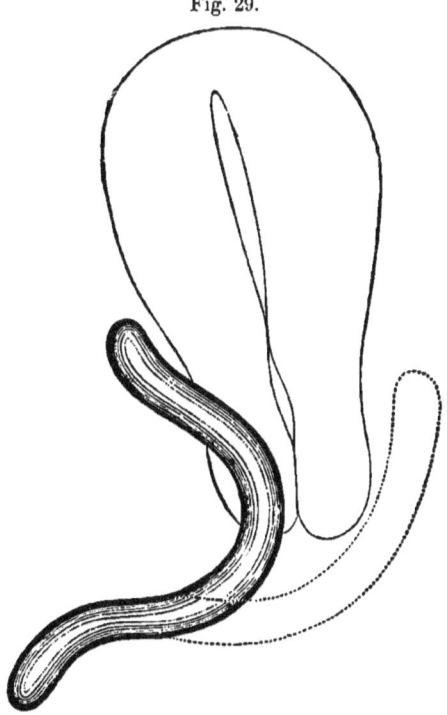

Fig. 29.

"1. Universal applicability to all cases in which any vaginal pessary can be retained, *i. e.*, where the perineum has not been destroyed by laceration.

"2. The perfect ease and comfort with which it may be worn.

"3. The ease with which the pessary is introduced and removed, even in virgins, without so much as introducing the finger into the vagina, going to its proper place with unerring certainty without the possibility of mistake, and hence,

"4. The perfect command the patient has over the instrument, allowing her to remove and replace it as often as comfort or cleanliness may require, thus avoiding to the fullest extent, the objectionable features of nearly all other pessaries.

"5. The facility with which the length or breadth or size of the curves may be changed either in the hard or soft rubber, so as to adapt it perfectly to the individual and to the progress of the cure.

Fig. 30.

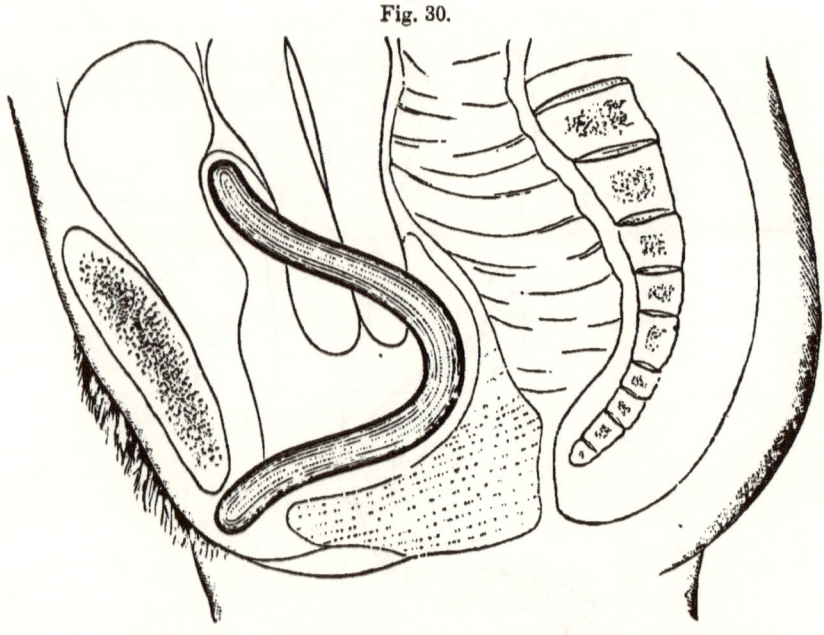

"6. The very slight tendency to displacement even in cases of extremely short vaginal cervix and extreme anteversion; since the lower segment, if properly adapted, will by its lever action, hold the uterine extremity well forward in contact with the bladder.

"7. The avoidance of distension of the vagina, thus maintaining its columnar character and allowing the greatest possible penetration of the instrument upward, in front of the womb.

"8. Its extreme simplicity of construction, being free from joints and complicated parts, and hence,

"9. Its cleanliness.

"10. Its cheapness, which is not excelled by any.

"Respecting the etiology of anteflexion, although considerable attention has been devoted to this subject and to elucidating the influence which areolar hyperplasia and general mal-nutrition may exert, it seems to me that sufficient importance has not yet been attached to *incomplete development* as a factor in these cases. It is to be borne in mind that prior to puberty the uterus is normally anteflexed to a marked degree and it is only when the organ undergoes the developmental change incident to the commencement of sexual activity, at puberty, that the womb erects itself into the position it should assume from this time forth.

"Thomas suggests that shortness of the round ligaments may prevent this erection of the womb, but so far, at least, as my observation extends, I have never been convinced that this is actually the case. Generally, indeed I should say *always*, the fundus may readily be carried up far enough by the uterine sound, and if it does not remain there, it is due to causes in the uterine tissues or to pressure from above. In some instances this tissue weakness may be, and doubtless is, a true areolar hyperplasia; but in many, I am convinced, it is rather a simple *puerility* of the organ. Menstruation may have been established but it is often imperfectly, and the whole sexual system seems to have but half reached perfection. The sexual instinct is generally wanting and quite often vaginismus takes its place, or at other times pruritus, or possibly nymphomania.

"It is in such patients that what Hewitt has called 'uterine lameness' is to be found; a condition which he ascribes to general muscular atony and mal-nutrition.

"It has seemed to me at least questionable whether the general depravity of nutrition, persistent anorexia and vegetative sluggishness were not rather due to the deranged innervation arising from the irritation set up in the reproductive organs. No surgeon can have failed to notice the profound effect sometimes produced upon an entire limb by some disturbance or injury of a painful but often not of a disabling character. It is quite as easy to understand how chronic uterine irritation might, through its universal nervous relationships, impair the digestive function and hence the nutrition. If this is so, it suggests the necessity of directing our attention to the uterine disease as the *cause* and not the *effect*, and while there can be no dispute as to the importance of improving nutrition by all available means, I am strongly of the opinion that this can only be done satisfactorily in proportion to the progress of the uterine treatment.

"Artificial supports will contribute very much to this, as will also stimulation of the reproductive system by the use of the galvanic or faradic sound, the application of these currents to the ovaries also, the regulation of the menses by suitable medication, and possibly the patient may be made to assist by a systematic course of massage applied to the breasts. These will generally be found ill-developed and flabby. Their sympathetic relations to the uterus should not be overlooked."

When the uterus cannot be replaced by the fingers and position alone, use the sound, or Whitney's, or Gardner's elevator.*

Medicinal Treatment.

The administration of those specific remedies which have the power to remove some of the morbid conditions upon which the flexions depend. According to many excellent authorities the chief cause of flexions is inflammation or congestion of the parenchyma of the uterus. If this condition predominates in the posterior wall, *retro*flexion occurs, if in the anterior, *ante*flexion. I believe this theory to be true in the large majority of cases. There are medicines which correspond to this caustive condition, the most important of which are *Sepia, Murex, Belladonna, Cimicifuga, Helonias, Lilium, Podophyllum, Æsculus, Platinum, Palladium,* and *Ustilago.* If

---

* I prefer Sim's Sound on account of its lightness, and also that it is less liable to cause irritation. The cuts opposite will show the difference.

UTERINE CAUSES. 141

Fig. 31.   Fig. 32.   Fig. 33.   Fig. 34.

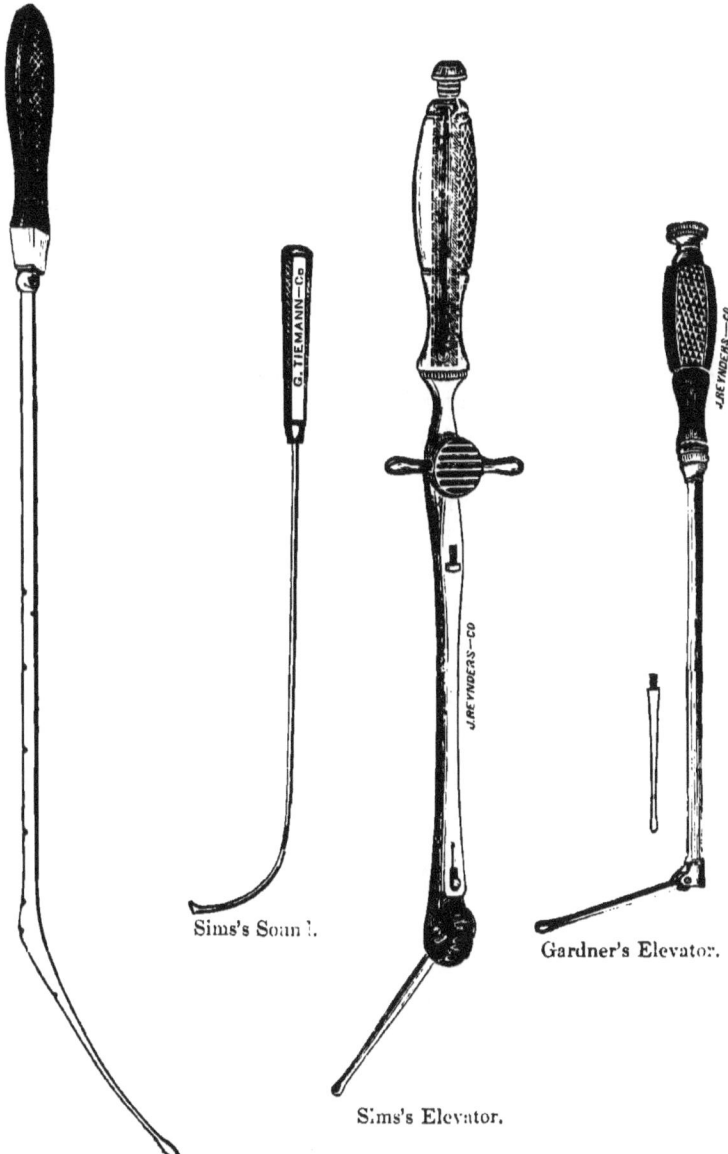

Simpson's Sound.   Sims's Sound.   Sims's Elevator.   Gardner's Elevator.

the retroflexion or version is of long standing, the aim of treatment is to remove the abnormal deposit in the parenchyma of the uterus, and impart tone to its weakened ligaments. This can generally be done by the administration of *Aletris, Hydrastis, Ferr. iod., Ferr. brom., Ignatia, Nux vomica,* and the use of an appropriate pessary.

In most cases the persevering use of Sitz-baths of a moderately cool temperature for fifteen minutes daily will greatly aid the other medicinal measures, as will also the use of the electrical current.

It will often occur that a sterility will be cured by the pessary alone *if it is worn during coition.* I observed this fact many years ago, before Sims's work was published. I have treated several ladies for retroversion, who objected to the use of the pessary for the reason that "if it was worn, pregnancy would result;" and several have preferred to endure the pain and trouble of a retroflexion than to wear a pessary, for this same reason.

Sims relates two cases of obstinate retroversion with sterility in which no sort of pessary should be worn except cotton; without the cotton pessary the uterus in each was turned back to an angle of more than 100° from a normal line; but with this pushed snugly up into the cul-de-sac the organ was comfortably sustained in position. Each of these patients conceived during the time of using this support. They were taught to apply the cotton pessary (a wad of cotton-wool of the proper size, with a small cord passed through the middle, 8 or 10 inches long) on rising in the morning, and to remove it on going to-bed at night.*

LATEROFLEXION of the uterus may cause sterility in the same manner as the before-mentioned displacements.

Inclination of the uterus to one or the other sides never attains a high degree; the fundus is inclined either to the right or left, whilst the vaginal portion takes the opposite direction.

---

* Uterine Surgery, p. 286.

Lateroversion is often combined with obliquity of the uterus. This displacement may be congenital or acquired. The latter form may be the results of metritis. I have observed that it occurs some time after labor or a miscarriage, when the woman persisted in lying all the time on one side. It may be caused by tumors in the body of the uterus, ovarian tumors, or disease of the round ligaments.

Treatment.—The medicines must be selected by the characteristic symptoms of the case. Mechanical treatment by pessaries is of doubtful utility. I think I benefited one case, and partially removed the flexion, by placing between the cervix and the vaginal wall an air-filled rubber disc pessary of about two inches diameter. The fundus was first pushed up as far as possible. The patient was directed to lie altogether on the side opposite to that occupied by the fundus uteri.*

PROLAPSUS UTERI.—Unless the prolapsus is attended by other morbid alterations in the uterine tissues, and is so complete as to constitute *procidentia*, it does not become a cause of sterility. A partial prolapsus may, indeed, favor impregnation, by allowing the spermatozoa a more ready entrance.

Whenever the cervix uteri passes through the mouth of the vagina we call it a procidentia, whether it be to a slight or a great extent. A procidentia may be complete or incomplete; complete, when the vagina is inverted and protruded externally; incomplete, when the cervix uteri alone projects between the labia for an inch or two, and remains thus stationary for a long time; usually it goes from bad to worse, till it eventually passes entirely through the vulva, forming a tumor of great size, which, at its most depending part, presents the os tincæ often ulcerated and bleeding. Sometimes the whole body of the uterus protrudes from the vagina; at others, the cervix, greatly hypertrophied and elongated, is the protruding portion.

Treatment.†—The medicinal treatment of prolapsus consists

---

* Studley's curved stem pessary might be of great service in these displacements.

† An excellent monograph on the medicinal treatment of prolapsus uteri has lately appeared from the pen of Dr. Eggert, of Indianapolis, Indiana.

in the administration of those remedies which will remove the pathological condition upon which it depends. This is generally *chronic congestion* of the uterus, and, unless of too long continuance, will yield to *Aurum, Belladonna, Sepia, Nux vomica, Platinum, Murex, Lilium, Cimicifuga, Aquaphobin, Ustilago, Secale, Podophyllum, Æsculus*, and *Helonias*.

No remedy or support will be of value unless the woman dress properly. The use of tight corsets and heavy skirts must be abandoned, for the pressure on the abdomen will prevent any remedial measures from effecting a cure. The *internal* support which I prefer to all others is the soft rubber elastic pessary of Dr. Jackson, in which the copper wire which is inside the rubber is open at the top and bottom from half an inch to an inch, the widest opening to be placed upward. It should be bent or moulded by the fingers into a shape approximating the retroversion pessary of Thomas or Hewitt: when in the vagina it opens of its own accord, and the top, in which the wire is left out for an inch, passes up behind the cervix, and affords a soft, elastic spring for the fundus to rest upon, and also allows the free passage of fæces along the rectum. The next best is the ordinary elastic ring, or oval pessary, which has been in use for many years.

Fig. 35.

Elastic Ring Pessary.

In *procidentia* it is evident that no medicinal means are of any value, unless the uterus can be replaced by pressure, and kept in position by simple mechanical appliances.

In *complete procidentia* surgical treatment should be resorted to. If the cervix is enlarged and elongated, the operation of amputation is quite successful.

Dr. A. K. Gardner reports several cases, in one of which the amputated cervix weighed nine ounces, two drachms and one scruple, "the largest on record as having been removed during life." He says: "The organ drew up far into the vagina after the portion was removed."*

Dr. Sims says:† "I amputate the cervix only when its lower

---

* Amputation of the Cervix Uteri.
† Uterine Surgery, p. 294.

segment is too large and too long, and projects so far into the vagina as to present a mechanical obstacle to the retention of the uterus *in situ* when replaced."

This operation may cure a sterility by allowing a fruitful coition, which is next to impossible when the cervix is hypertrophied, elongated, and protruding.

The so-called *prolapsus uteri, without sinking of the fundus*, to which Virchow has called attention, may act as a cause of *sterility*.

In this variety there is complete inversion of the vagina, with the external orifice of the uterus situated at its extreme end. The enormous elongation of the uterus affects chiefly the cervix, the substance of which becomes dense and vascular. In this condition of hypertrophy and elongation of the cervix, the uterine cavity is sometimes narrowed, or atresia of the internal orifice may be found.

The indications for treatment, aside from the elevation of the uterus by mechanical supports, are to dilate the internal orifice by bougies or an operation, after which conception will readily occur if no other obstacle is present.

Dr. Sims's operation for the cure of complete procidentia uteri consists in *removing a portion of the anterior wall of the vagina*, bringing the edges together, and fastening them by metallic sutures. This simple operation seems to prevent the return of the uterus to its abnormal position. For a complete history of this successful operation see Sims's *Uterine Surgery*. It seems evident that this operation is vastly superior to the *perineal* operation as performed by Mr. Baker Brown, Dr. Savage, and others.

In cases where this or any operation is objected to, we may do something to alleviate the condition of the patient, and, *perhaps*, bring the uterus into such a position as to be capable of being impregnated. Dr. Sims says that in such cases no pessary, except Zwang's, is of the slightest value; while the best support is simply a large tampon of cotton, wet with glycerin, which may be introduced in the morning, to be worn all day.

It must be remembered that impregnation will occur whenever the seminal fluid is brought in contact with the cavity of the uterus, if the connection between that organ and the ovaries is not disturbed. If the uterus is so far protruding that the ovule cannot pass from the ovary into its cavity, conception cannot take place, even if the semen is thrown into its cavity.

ELEVATION OF THE UTERUS.—This displacement, together with its attending organic changes, may cause sterility.

By *elevation* of the uterus is meant a displacement of the entire organ upward. The fundus ascends into the peritoneal cavity, both peritoneal cavities thereby becoming shallow, and the fornix of the vagina being so stretched as to cause it to become *cone-shaped*. At the same time, the vaginal portion disappears more or less, leaving merely a button-shaped rudiment. (Rokitansky.)

The causes of elevation of the uterus lie either within or are external to the organ itself. The causes that originate in the organ itself are: an increase in volume of body and fundus, unless the latter has been previously bound down by adhesions in the pelvic cavity, in consequence of which, ascent of the uterus would be hindered; distension of the uterine cavity by mucus or blood; formation of fibrous tumors, etc.

To the class of external causes belong tumors of the broad ligaments or ovaries, adhesions formed during pregnancy or the puerperal state, previous to complete involution; also, vascular tumors of the vagina, tumors of the pelvis, etc.

Slight degrees of elevation are found combined with flexions, and always with anteflexion.

The effect of the elevation of the uterus, especially when produced by other causes than tumors pressing from below, is *elongation* of the organ, *chiefly of the cervix*, to sometimes twice or thrice its normal length; this elongation being generally accompanied with diminution of its cavity and thinning of its walls. The diminution of the uterine cavity by longitudinal traction, is always more considerable near the internal orifice. Even complete occlusion of the canal may take place. In isolated cases, an obliteration of the canal at the above point occurs in consequence of rupture of so-called Nabothian glands, from the ruptured walls of which connective tissue is produced, which ultimately leads to complete atresia of the canal by agglutination. (Kolb.)

Mere elevation of the uterus will not cause sterility, for the spermatozoa find no obstacle in distance, if the mucous membrane is healthy and the cervical canal open. But if occlusion or atresia occur with the elevation, sterility surely results.

Treatment.—The removal of the causes of the elevation by medicinal or surgical means constitute the indications for treatment. If occlusion or atresia is present, the canal should be opened by the operation recommended and practiced by Sims. After this is done, conception will occur as readily as if the uterus was low in the pelvis. If large tumors of the pelvis or uterus caused the elevation, their removal by medicinal and surgical treatment should be attempted. If this is impossible, and the tumors are so large as to offer a probable impediment to childbirth, it is proper to advise the avoidance of pregnancy, or the operation referred to, unless a retention of the menses, or a dysmenorrhœa, render it necessary.

INVERSION OF THE UTERUS.—This dangerous and infrequent form of displacement consists in the turning of the uterus inside out. As the bottom of a bag may be pushed through its mouth so that the inner surface becomes the outer, so may that of the uterus. Inversion may be *partial* and *complete;* partial when the body has become depressed, but has not passed through the os; complete, when the uterus has been completely turned inside out, and the inverted fundus and body hang in the vagina or between the thighs. Inversion of the uterus generally occurs shortly after labor, when the uterine tissues are relaxed. At such a time, undue traction on the cord, to expedite the birth of the placenta, coughing, sneezing, change of posture, straining at stool, etc., may cause the accident.

It is possible that inversion may occur at other times; and there is evidence on record that a non-pregnant and undilated uterus may become inverted.*

---

* See Thomas's Diseases of Women, p. 339.

As this displacement may become *chronic*, it should be placed among the causes of *sterility*, although Gardner does not mention it. Chronic inversion may continue for many years, giving very little annoyance; or it may render the life of the patient miserable, on account of hæmorrhage and other attending symptoms, and nevertheless last for years.

DIAGNOSIS.—Inversion may be mistaken for a polypus or a fibroid tumor. For accurate differentiation of this displacement from tumors, reference is made to Thomas, Sims, Guernsey, Lee, and other authorities:

Treatment.—There are three plans of treatment:
1. To return the uterus to its place.
2. To leave it displaced, and adopt means preventive of hæmorrhage.
.3. To remove it by amputation.

For the cure of sterility depending on this displacement, the *first* plan is the only one to be adopted.

In chronic cases, the second and third plans were considered the only means possible; but it has been demonstrated by Dr. Tyler Smith, of London, and Dr. White, of Buffalo, that even after the continuance of inversion for ten or fifteen years, reduction is possible. Since these successful attempts, the operation has been performed by Sims, Thomas, and many other obstetric surgeons.

The diagnosis having been clearly made, and reduction determined upon, the bowels and bladder should be emptied, and the patient put under the influence of an anæsthetic, and laid on her back on a strong table. The operator should always be attended by three or four reliable counsellors, upon whom he may call, not only for advice but physical aid. Having thoroughly oiled one hand, the nails of which have been pared, he should now slowly dilate the vagina so as to introduce it, and grasp in its palm the entire tumor. The other hand should be laid upon the abdomen, so as to press just over the ring which marks the non-inverted cervix, and oppose the force exerted through the vagina, so as to prevent too great stretching of this canal. In some cases a conical boxwood plug has been used.* For other methods of reducing obstinate inversion, consult the authorities above named.

After the reduction is effected, the patient should remain in bed, with the hips elevated, for a week or more, and the remedies indicated by the symptoms should be administered.

(Consult—*Aletris, Belladonna, Caulophyllin, Cimicifuga, Conium, Ferrum, Helonine. Ignatia, Mercurius, Nux vom., Podophyllin, Ustilago, Lilium, and Sepia.*)

TUMORS OF THE UTERUS.—Sterility may be caused by any of the varieties of abnormal growths known as tumors of the uterus. The most common are the polypi, and those denominated *fibrous*. They prevent conception, either by blocking up the canal of the cervix, the cavity of the uterus, or the orifices of the Fallopian tubes.

POLYPI.—Dr. Sims† describes the location and character of these polypi as follows:
1st. Those growing from or about the os tincæ.
2d. Those growing in the canal of the cervix.
3d. Those growing in the cavity of the uterus.

The first may be fibro-cellular or mucous. The second are almost always mucous. The third are almost always fibrous.

FIBROUS.—The uterus is particularly prone to the development of firoid tumors. They frequently prevent conception, even when located in the parenchyma of the womb, but not necessarily so. They are classed according to the manner of their attachment to the walls of the uterus, as extrauterine, intrauterine, and intramural.

EXTRAUTERINE fibroids, grow from any portion of the external surface of the uterus, and may be pedunculated, or they may be sessile, with a broad, immovable attachment to its outer muscular tissue.

---
* See Thomas's Diseases of Women, p. 345,
† Uterine Surgery, p. 67.

The INTRAUTERINE project into the cavity of the womb, and, like the first, may be pedunculated or sessile.

The INTRAMURAL are so called because they are imbedded in the walls of the uterus, being interlaced and overlapped in all directions by its muscular fibres.

By reference to the statistics it will be seen that fibroid tumors of the uterus were the causes of sterility in 95 out of 505 cases.

Treatment.—The surgical treatment of all polypi may be summed up as follows: (*a*) Excision. (*b*) Torsion. (*c*) Ligature. (*d*) Ecrasement. (*e*) Pressure by the sponge-tent.

Small polypi may be excised by the knife or scissors, if the pedicle can be reached; but large growths are apt to bleed profusely, and even dangerously, after the operation.

Dr. Sims, however, regards the hæmorrhage as easily managed by the use of *perchloride* or *persulphate of iron*.

Small, and even large tumors, may be twisted off with the forceps, and the danger from bleeding is less than when excision has been resorted to. In intrauterine polypi, when it is necessary to expel them through the os, Ergot is generally used. If this agent fails, I can favorably recommend the introduction of slippery elm bougies, between the tumor and the uterine walls, for that purpose.

(In a case occurring in my own practice, a small mucous polypi, attached to the canal of the cervix, near the os externum, was removed by one application of the solid nitrate to the pedicle).*

The ligature, once so popular, is now rarely employed. Ecrasement constitutes the safest and most expeditious of all the operations.

Dr. Simpson has used the sponge-tent with singular success in the treatment of polypi.

For minute directions for operating on polypi, refer to Sims's *Uterine Surgery*, Thomas on *Diseases of Women*, Baker Brown on *Surgical Diseases of Women*, and Franklin's and Helmuth's works on *Surgery*. No homœopathic physician should neglect

---

* See United States Medical and Surgical Journal, vol. ii, p. 235.

to place these valuable surgical works in his library, for cases may arise where his reputation and that of our school in the locality of his practice may depend on his successful surgical treatment of uterine tumors.

The *medicinal* treatment of polypi is not mentioned in allopathic works, because no medicines are believed by them to exert any curative action on such tumors.

"The cure of polypus by medicinal agency," says Leadam,* "cannot be accomplished by the old school of medicine; . . . . . . but homœopathy has rendered results which prove the curative power of its remedies."

Staphisagria and *Calcarea* are those which have been recommended principally to remove the state of dyscrasia upon which the formation of polypi depends. The remedies which are pointed out by Jahr† are too numerous to be quoted.

Guernsey‡ gives the *characteristic indications* in his peculiar manner for the following medicines: *Aurum, Calc. carb., Conium, Lyc., Mercurius, Mezereum, Nitric acid, Petroleum, Phosphorus, Phosphoric acid, Platina, Pulsatilla, Silicia, Staphisagaria, Teucrium*, and *Thuja*.

These remedies are also recommended for *all* varieties of tumors; but as the indications are mainly theoretical, but few clinical cases having verified them, we may consider their value in polypi, or fibroid tumors, as quite problematical.

The surgical treatment of *fibroid* tumors is thus summed up by Thomas,§ who seems to have consulted almost every surgical work.

(*a*) Absorption. (*b*) Excision. (*c*) Ecrasement. (*d*) Enucleation. (*e*) Sloughing. (*f*) Gastrotomy. Thomas doubts if their absorption can be excited by any medicine (allopathic). Scanzoni says he does not know of a single case where medicines caused a complete cure. Sims believes that pressure by means of sponge-tent or sea-tangle, may cause absorption. The treatment by excision or the ecraseur

---

* Diseases of Women, p. 274.
† Ibid., p. 224.
‡ Obstetrics, p. 99.
§ Diseases of Women, p. 407.

is fully described by the last-named author. Enucleation has been successfully practiced by Amussat, Atlee, West, Ludlam, Danforth, Adams, Helmuth, Franklin, Beckwith, and others. Sloughing, once recommended by Dr. Baker Brown, has been abandoned. Gastrotomy, or removal of the tumors, and with them the uterus, is a rare and dangerous operation. According to Dr. H. R. Storer, who has written the history of such operations, there have been 24 operations, with 18 deaths. The *medicinal* treatment of fibroids is not very satisfactory. In allopathic practice, Dr. Channing, of Boston, claims to have cured many by internal medication. Dr. Simpson seems to have great faith in the long-continued use of *bromide of potassium*. Dr. Sims says he has never seen the slightest effect produced on such tumors by internal medication. In homœopathic practice no clinical experience has yet been published showing the effects of our remedies in this disease. Guernsey recommends the same medicines as for polypi. I have a case of intramural fibroid tumor which appears to have slightly diminished in size under the use of *iodide of arsenic* 3d, continued for two months. I would suggest the protracted use of *hydrastis* and *turpentine*. Within a few years the therapeutic measures for the treatment of *uterine fibroids* have been greatly increased by the discovery of the curative power of the persistent use of *ergot*. It has been found that, in a large proportion of cases, the tumors decreased under its use, and in many cases they nearly or altogether disappeared.

Ergot, to be of value in such tumors, must be administered in large doses, large enough to institute and keep up its primary action, viz., *a contraction of the uterine arteries* and the uterine muscular fibres. By this action it cuts off the *nutrition* of these tumors, and they cease to grow, or shrivel away, because they are no longer fed by the blood.

The *Bromides* undoubtedly act in the same manner. Also *Ustilago*, which is a very near relation of ergot. If either Ergot or Ustilago, is given by the mouth, the dose must be from 15 to 60 drops three times a day. But sometimes the stomach will not tolerate ergot, and it must be administered by *hypodermic injection*. From 10 to 30 drops of Squibb's aqueous extract, or an aqueous solution of *ergotin*, is injected into the cellular tissues of the abdomen or thigh, once a day. Hildebrand injected about 3 grains at one injection, daily, of the aqueous ext. His formula was, 3 parts aqueous ext. Ergot, 7½ of Glycerin, 7½ of Water. Ustilago might be used in the same manner, and in the same doses, if we had an aqueous extract.

The *Iodide of barium* is a powerful remedy against abnormal growths, and may become useful in removing uterine fibroids. The continuous current from a powerful galvanic battery of 20 to 60 cells has been found effectual.

Diet in the treatment of uterine fibroids.—In a recent number of the *American Journal of Obstetrics* (October, 1877), Dr. E. Cutter has a very interesting paper on the treatment of uterine fibroids by means of a peculiar diet. He says he got the idea from Dr. Salisbury, of Cleveland, who regards these growths as generally due to excess of carbohydrates, starches and sugar, fermentable food, in the diet; that they are usually caused by disorders of nutrition, and that by feeding patients on a diet composed of animal food, the condition which was most active in bringing on the diseased condition is removed, and the system is enabled to right itself by its own recuperative power. Dr. Cutter reports *eight* cases in which women with uterine fibroids were placed on such diet; all were greatly improved and some were cured. The following were the articles permitted and prohibited:

PERMITTED.	PROHIBITED.
Beef in all forms.	Starches and sugars.
Tripe, veal.	Common white flour, in all
Calves' foot and head.	and every form.
Pork, fresh, salt and cured.	*Bread, biscuits, crackers.*
Sausages and ham.	Cakes, doughnuts, puddings.
Mutton.	Rice.
All kinds of game.	Potatoes in any shape.
Milk, butter, eggs.	*Sugar*, candy.
Cheese, cream.	Corn-starch, arrowroot, etc.
Vegetables, without or with little starch.	
Fish, fresh and salt.	

If the strict diet becomes very repulsive, a mixed diet can be allowed for a short time, of *whole wheat, Graham bread, oatmeal, rye-meal,* and *corn-meal.*

INFLAMMATION OF THE UTERUS.—Acute and chronic inflammation of the uterus usually result in sterility; not only by the changes produced in the density of the tissues affected, but in the changes effected in the secretions from the diseased surfaces.

The inflammatory affections of the uterus causing sterility are:

(1.) *Endometritis* (inflammation of the lining membrane of the uterus). (2.) *Endocervicitis* (inflammation of the lining membrane of the cervix).

ENDOMETRITIS has been described under the names of uterine catarrh, uterine leucorrhœa, and internal metritis. Its location extends from the os internum to all portions of the uterine cavity, or fundus. (A variety has been described by Dr. Routh, as *fundal* endometritis. In this variety, the inflammation is located in the fundus, occupying only a small spot.)

As endometritis is a frequent cause of sterility, it will be well to devote a space to its consideration. Its chief symptoms are: (*a*) *Leucorrhœa.* (*b*) *Menstrual disorders.* (*c*) *Pain in back, groins, and hypogastrium.* (*d*) *Nervous disorders.* (*e*) *Tympanites.* (*f*) *Symptoms of pregnancy.* (*g*) *Sterility.*

Treatment.—No mention is specially made of endometritis by Marcy and Hunt, Leadam, Gollman, or Mintern. Jahr* treats of it in his own peculiar manner. Ludlam and Guernsey† give the only good description we have in any work in our school. Jahr's remedies are: *Alumina, Calcarea, Kreosotum, Mercurius, Pulsatilla, Sepia, etc.,* for which he gives the symptomatic indications. Guernsey gives the characteristic or key symptoms of seventy-two remedies, among which, *Arsenicum, Borista, Calcarea, Conium, Graphites, Lachesis, Kali bi., Kali hyd., Mercurius, Muriatic acid, Nitric acid, Platina, Podophyllum, Pulsatilla, Sabina,*

---

* Diseases of Women.
† Obstetrics, p. 128.

Sepia, Sulphur, and *Zinc,* are the most useful, and especially adapted to the chronic form of the disease.

In addition to the above medicines, I have found *Cimicifuga, Gelsemium, Hamamelis, Helonias, Hydrastis, Myrica, Lilium, Phytolacca, Senecio, Trillium,* and *Xanthoxylum,* useful in some cases.

CIMICIFUGA is especially indicated for the nervous symptoms accompanying the disease, also when the pains in back, loins, etc., are present.

GELSEMIUM is superior to all others in the acute or subacute stage.

HAMAMELIS when the discharge is "rusty," or the menses continue nearly through the month; (also *Trillium, Erigeron,* and *Senecio.*)

HELONIAS for chronic albuminous leucorrhœa (uterine), with great debility.

MYRICA when the discharges are offensive, bloody, and corrosive: (also *Arsen. iod*).

PHYTOLACCA for nearly the same symptoms, especially if syphilis is present.

HYDRASTIS if the discharge is ropy, thick, yellow and tenacious, with great debility. There are many points of resemblance between these and the older remedies, which may be studied to advantage.

I believe I have cured cases of *Sterility* from endometritis by means of the persevering use of *Calcarea, Conium, Mercurius, Nitric acid, Sabina,* and *Sepia;* also *Cimicifuga, Helonias, Lilium, Phytolacca,* and *Hydrastis.* Some of these remedies I have used topically, by injecting a solution into the cavity of the womb. One case, in particular, is worthy of mention: A lady who had borne one child, had been sterile seven years. She had the most profuse, thick, tenacious leucorrhœa I have ever seen. The vagina was also affected with chronic inflammation, and the os abraded. The two latter were cured; the former by injections of *Hydrastis,* the latter by a few applications of *Nitrate of silver.* The uterine discharge, however, still continued; a thick rope of mucus constantly hung from the os. With a small syringe, having a long, slender tube, I injected, once a week, ℨij of a solution of *Muriate of hydrastia** into the cavity of the uterus (the cervical canal allowing its return). After six injections she became pregnant, and went her full time.

Molesworth's is the safest and best intrauterine syringe, for it allows the fluid to return from the cavity.

Fig. 36.

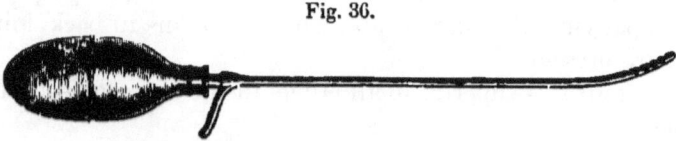

Molesworth's double canula and bulb syringe.

Another instrument, quite as safe, but less convenient perhaps, is Nott's double canula catheter, through which the medicated injection is thrown by an ordinary syringe. The arrows denote the course of the current.

Fig. 37.

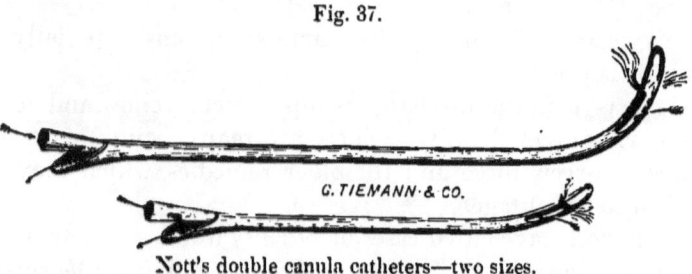

Nott's double canula catheters—two sizes.

The remarks of Sims on the surgical treatment of this disease are so important that I quote them: "The first great principle to guide us is that of insuring a very free exit from the cavity of the uterus for the secretions therein generated. The second is that of appropriate local applications to this cavity, for the purpose of modifying or healing, as it were, the diseased surfaces. When the canal of the cervix is contracted, I have freely divided it, as in cases of dysmenorrhœa dependent upon mechanical obstruction, and this with great relief. The uterine secretions must not remain pent up in its cavity. With a patulous cervix, one may use medicated injections, etc." For

---

* ℞. Muriate of hydrastia, gr. v; Glycerin, ℨss.; Water, ℨiss.

intrauterine applications, use small cotton tampons, to which a string is attached, medicated with one of the following preparations, and applied by means of my intrauterine applicator, or Wylie's cervical protector.

Fig. 38.

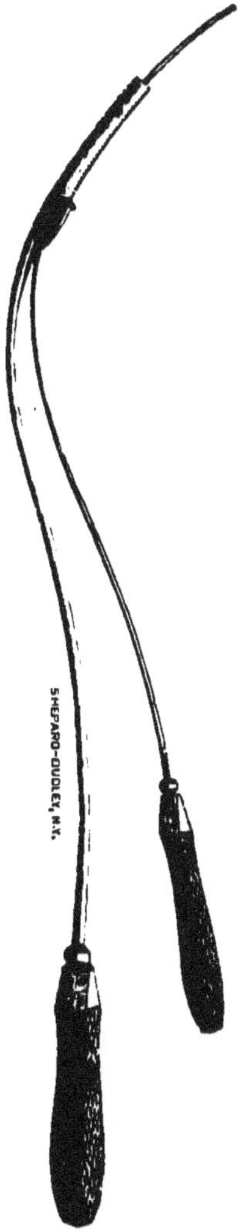

In using Wylie's instrument, wrap cotton around the rod, medicate it, and insert it into the tube, which has been passed into and through the cervical canal. Push the rod up to the fundus and move it from side to side a few times, then withdraw both rod and tube.

For these intrauterine applications I have found the following the most efficacious:

Merrill's fluid hydrastis,	ʒj	to Glycerin	℥j.
Sulphate of hydrastia,	gr. x	"	"
Muriate of "	gr. v	"	"
Tinc. calendula,	ʒj	"	"
" thuja,	ʒj	"	"
" grindelia,	ʒij	"	"
" cubebs,	ʒj	"	"
" copaiva,	ʒj	"	"
Carbolic acid,	ʒj	"	"
Kali bich.,	gr. v	"	"

Wylie's cervical protector.

My applicator can be used in the same manner, and is a much cheaper and more convenient instrument.

Dr. Sims has met with several cases of *fundal* endometritis, and thus gives its diagnosis: "Place the patient in the left lateral semi-prone position; introduce the speculum; hook a tenaculum slightly into the anterior lip of the os tincæ; draw this gently forwards, pulling the os open, so as to be able to look right into it; then pass the sound, previously warmed, gently along the cervix, using no force whatever, but almost letting it go by its own gravity, as it were, to the fundus. This is attended by no pain whatever till the sensation point be reached, when it produces the most intense agony—a pain that does not cease sometimes for hours after the experiment."

In this disease there is little or no discharge, but generally pain in one or the other hip, generally the left, in the inguinal region, groin, ovary, and even mammary region.

Dr. Sims relates one very severe case, which he cured in a few weeks, by dilating the cervix, and simply injecting into the cavity of the uterus *a few drops of glycerin,* twice or three times a week. In the course of a year, this patient became a mother, and has had three children since.

In another similar case, he says: "A single sponge-tent, followed by the injection of half a drachm of the officinal *tincture of iodine* produced almost complete relief at once. A repetition of the same, ten or twelve days afterwards, produced a perfect cure." Medicated injections into the cavity of the uterus are not always safe, even if the cervical canal is dilated.

There are *four* other methods of applying medicinal substances to the internal cavity of the uterus (the endometrium), which are preferable to the use of injections, namely:

(*a*.) The *brush* or *swab*. The brush is made of pig's bristles, and should be not over one-fourth inch in diameter. The *swab* is simply a bit of "absorbent cotton-wool," or clean, white cotton-wool, wound around the end of a flat probe or applicator to the length of one or two inches, and one-quarter or one-eighth inch thick. The cervix should be previously dilated. When the *brush* or *swab* is used it should be passed quickly up to the fundus uteri, and moved from side to side or rotated, so as to bring it in contact with the whole surface of the cavity.

(b.) The *cloth-tent*, first introduced by Dr. Taliaferro, of Columbus, Georgia, and into our school by Dr. E. W. Beebe, of Wisconsin,* although they have been used for centuries, even by Hippocrates. The following is the method of their manufacture, as described by Dr. C. Leonard:

"To make one, you need but a strip of linen, six inches in length by three-quarters of an inch in width, a piece of hair wire four inches long, and a few inches of common thread. Roll one corner of the linen strip lightly between the thumb and finger, then unroll and place the centre of the wire at the corner so rolled, and then roll the cloth at this corner over it (*spirally*, just as you would go to work to make a paper lamp-lighter), till you get *almost* to the other corner of the same end, then bend the wire upon itself (double it, in other words) so that the two extremities will point to the unwound portion of the linen; this done, continue rolling the linen, in a *spiral* manner, about the doubled wire till exhausted, then tie with the thread the last spiral turn about the wire. You now have a tent about two and a half inches in length, and one sufficiently firm to enter *any* normal uterine canal, and almost any abnormal one. You can bend it to any curve you choose to facilitate its introduction.

"It has still another advantage over all other tents, in that you can leave it *in situ* (as I frequently do, for twenty-four hours) with *no* danger to your patient, as it is *inexpansible*, and hence no excitor of metritis, though a stimulator (from its very slight mechanical irritation) to the endometrium. By so doing you can get a *prolonged* action of a medicament upon the lining membrane of the uterus, which is impossible to get by any other method of application. Further, you need not use such energetic local applications, and you may be sure that they reach the *whole* uterine cavity; something you cannot do with our intrauterine applicators, unless you are a very skillful manipulator. The shape of the fundus cavity is an anatomical

---

* American Homœopathist, November, 1877 (Dr. E. W. Beebe).

proof of the great difficulty of making a complete application with the common metal applicators; whereas the cloth-tent, by meeting with resistance at the fundus, immediately doubles upon itself, thus occupying the whole cavity.

"You can make them of any *size*, and of any degree of stiffness, by increasing the thickness of cloth, and the size or number of doublings of your wire. I have them of all sizes, from those suitable for an ante-puberal uterus to one as large as your index-finger.

"I use them now for cleansing the uterine tract previous to an application of astringents or other medicaments thereto, and find they clean away the tenacious mucus much better than a syringe or a wisp of cotton on Emmet's applicator. Indeed, it is invaluable in many ways. By leaving the thread without the vulva, the patient can as easily and safely remove it at her residence as can her physician. You have only to remember to tie a string (or a colored thread) to the cotton pledget you leave in the vagina, so that she may be made aware which to remove (pull) first."

Dr. E. W. Beebe claims for them substantially the same advantages as those above mentioned. I have frequently used them, made extempore, without the wire, when the cervical canal is open enough to permit their introduction, or after I have dilated the passage by means of laminaria tents.

(*c.*) *Packing.*—This method has been used by many gynæcologists and by myself for years. It has lately been highly recommended by Dr. Carr,* of Galesburg, Ill. It consists of taking narrow strips of clean white linen or cotton, one-quarter inch by six or ten inches long. Saturate this in the medicinal solution, and folding one end over an applicator, push it gently up to the fundus; then, withdrawing the applicator or probe, push up another portion, and so on until only a short portion is left hanging in the vagina. It can be removed in a few hours.

(*d.*) *Slippery elm bougies*, made of the proper size, and soaked in the appropriate medicinal solution, or powdered over with the drug.

---

* American Homœopathist, February, 1878.

In all cases the drug selected should be that one which corresponds in its topical action to the diseased condition of the mucous membrane. This method is as strictly according to the law of *similia* as if the drug was given internally by the mouth. The curative action exerted by these local applications is "homœopathic," "alterative," or "substitutive," as we prefer to call it.

(*e.*) *The pith of the dried corn-stalk.*—Dr. W. T. Goldsmith, of Atlanta, brings this substance to notice in the *Transactions of the Medical Association of Georgia*, 1878. Take a joint of dried corn-stalk; strip it of its cuticle, and compress the pith, slowly and firmly, between the thumb and index-finger. By continued pressure, it is reduced four or five times less than its original size. It has a dilating power equal to sea-tangle or sponge. The corn-stalk tent is of easy introduction. Its rigidity overcomes any slight resistance. Dr. Goldsmith has used this tent for the last seven years. He has not had a single accident from its use, although he has introduced the tent many hundreds of times. The advantages of this corn-stalk tent are:

It dilates effectually, but not too rapidly.

It is smooth, soft, and can be removed without force.

It produces no lacerations, abrasions, or irritation of the mucous membrane.

It can be medicated with any substance as easily as the sponge or cloth tent.

It is of vegetable origin, and hence, does not become putrid and poisonous to the patient.

It may be retained, non-compressed for days, without injurious results, if no pain occurs.

A number of small tents, filling up the cervical canal, may be used for more rapid expansion.

It can be prepared in a few minutes of any desired curve, size and length.

Any degree of compression may be given it, or it may be used without compression.

It may be perforated, like the sea tangle, and its power of absorption increased, by packing its surface.

The following are the most useful agents for the above methods:

Comp. tinc. iodine, Tinc. ferr. mur., Conc. sol. carbolic acid, Ext. pinus canadensis, Nitric acid, Chromic acid, and all those mentioned in the list given above.

Those mentioned in both lists, all but nitric and chromic acids, can, in old, intractable cases, be used diluted *one-half glycerin*. Nitric and chromic acids should be diluted one-half with pure water.

The bougies should be introduced up to the fundus (first dilating the cervix with a sponge or laminaria tent if necessary), and allowed to remain there three or six, and even twelve hours, if they can be borne. If they cause continued pain, the patient should be told to remove them. They should be repeated every four or six days. It rarely requires more than two or three months to cure the most obstinate case, and conception can occur as soon as the mucous membrane is restored to a healthy condition. If the medicated bougie can not be borne, suppositories medicated with the same remedies act well; they soon melt and cause little or no irritation.

I have been successful with *medicated suppositories*, after the plan of Simpson and Sims. These are an inch and a quarter long, and small enough to pass along the cervix, and medicated with various remedies, so as to bring these into permanent contact with the diseased surface. I know that suppositories medicated with *Conium, Calendula, Hydrastis, Iodine, Sanguinaria, Iodoform, Carbolic acid, Tannin, etc.*, act very favorably in cervical induration or hypertrophy. The best forceps to use in introducing these suppositories into the cervical canal, or dressings to the os, is Bozeman's

Suppositories, for intrauterine use, are made of cocoa-butter or gelatin, with which are incorporated certain medicinal agents. The chief objection to their use is that they are very brittle and difficult of introduction, unless the cervical canal is open and straight. I have lately been using *flexible bougies* made of gelatin, which melt in the uterine cavity in an hour or two, and are much easier of introduction.* My favorite formula for the suppositories or bougies are, to each one, one-eighth inch thick and two inches long, Carbolic acid, one grain;

---

* They can be ordered of A. Arend, 179 Madison St., Chicago.

Fig. 39.

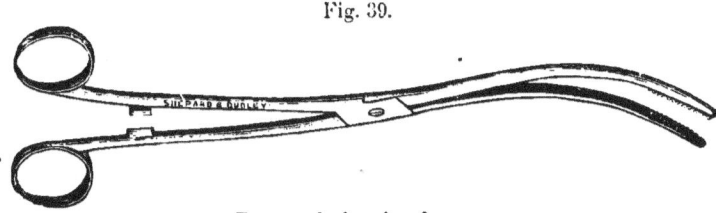

Bozeman's dressing forceps.

or Muriate of hydrastia, two grains; or Iodoform, two grains; or solidified Copaiva, two grains. These are pushed up with dressing forceps, and a cotton tampon placed against the os to prevent their extrusion.

In many cases of chronic *endometritis*, or *endocervicitis*, the diseased surface of the uterus becomes the seat of *fungoid growths*, *cystic degeneration*, or small, numerous *mucous polypi*. The discharge from these surfaces becomes bloody, "rusty," sanious, excoriating, or purulent. In such cases, all the methods of treatment I have mentioned may prove useless, or, at most, palliative, and the patient becomes the subject of frequent relapses. Of course, so long as this condition remains, *sterility* will remain.

There is, however, one method which we owe to the boldness of Sims and Thomas; a method which, notwithstanding its seeming severity, is painless and harmless if used properly. I allude to the use of the *curette* and *scoop*.

The simplest form of *curette* is a looped wire about the size of a large knitting needle, somewhat flattened. The loop itself is of an oval shape and a half inch long by one-third inch wide. This is attached to a handle—the whole instrument is about ten inches in length. The looped end is slightly bent at a long curve. When this instrument is used, the cervical canal should first be dilated so that it will easily enter. Push it up to the fundus and draw it gently back, at the same time pressing it against the uterine walls. It also may be pushed up from the cervix to the fundus against the walls. The effect of this movement is to push or pull off the fungoid granulations. Little or no hæmorrhage ever attends its use, and the patient rarely feels it unpleasantly.

Fig. 40.

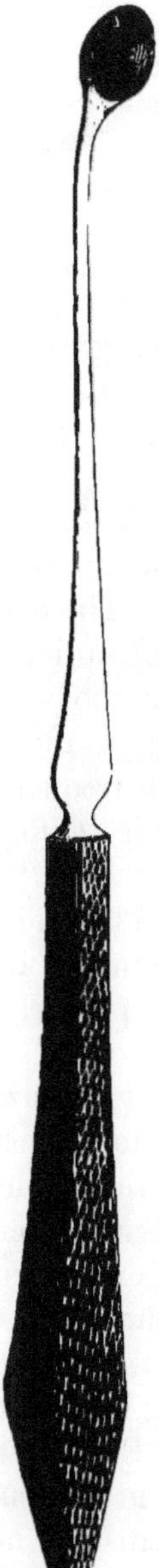

There are other curettes of various sizes and varying shapes, with *cutting* edges. These are to be used in severer cases, when the fungoid growths are large, or mucous polypi exist, or in cases of cystic degeneration.

They are used in the same manner as the wire curette, carefully but thoroughly going over every portion of the internal uterus. In some cases the amount of diseased matter detached and extracted is surprising—half an ounce or more—and the rapid recovery following their use is equally surprising. The removal of the morbid growths is always followed by a disappearance of the menorrhagia, leucorrhœa, uterine tenderness, and the general debility.

Fig. 41.

Another form of this instrument is called the *scoop*. They are of various sizes and shapes; some have a bowl like a teaspoon; others a round cup-shaped bowl. In some the edge is cutting, others dull, and still others serrated. They can be used instead of the curette, where that does not seem efficient; but they are generally used and recommended by T. G. Thomas, especially for the removal of sub-mucous fibroids, intrauterine tumors of all kinds, or when any foreign substance is adherent to or growing from the inner surface of the uterus.

I have operated many times with each of these instruments. At first I had doubts of their safety, but after some experience did not hesitate to use them. I should not like to practice Gynæcology without them, and I have never seen any ill-effects follow their use, but in every case *recovery*.

Sieman's scoop.

Thomas's serrated curette.

INTRAUTERINE FLEXIBLE PENCILS.—During the last year I have been using with the happiest results the medicated intrauterine flexible pencils, lately introduced into gynæcological practice. I believe I was the first to use them in the West, and they were first made in Chicago, at my suggestion, by Mr. Arend whose specialty is in that direction. They are now extensively manufactured by C. L. Mitchell, of Philadelphia. Their special sphere of usefulness is in the treatment of *endocervicitis* and *endometritis, uterine catarrh, mucous cervicitis*, etc. They are about three inches in length and one-eighth of an inch in diameter, and are composed of gelatine and glycerin, with which are incorporated any remedy indicated by the nature of the case. The first I used were medicated with *Muriate of Hydrastis*. I have since used the *Iodoform*, which are of the greatest value in the worst cases. Next to them I prefer pencils of *Carbolic acid, Gelsemium, Belladonna, Calendula, Hamamelis, Grindelia, Ergotine* and *Tannin*. Those who never used them have no idea of the facility with which they can be introduced and their vast superiority over the solid, unyielding crayons or pencils. They will slide into the cavity of the uterus in a cervical canal through which it is difficult to pass the smallest sound. By virtue of their flexibility they readily pass through a *tortuous* canal. Unlike a stiff pencil they do not catch in the crypts of the cervical canal. They melt slowly, requiring probably several hours, so that the medicinal substance is fully distributed over the diseased surface, and remains long enough in contact to excite their specific influence. They are introduced through a speculum by means of a slender dressing forceps, and pushed up *within the os*. It is best to place a ball of cotton moistened with dilute glycerin *against* the os. Sometimes it is difficult to push the pencil within the os with the forceps, or to withdraw the forceps leaving the pencil within. To obviate this I had a simple instrument made which answers a good purpose. Upon the end of a stiff wire is placed a small cup, one-quarter of an inch deep, and with a diameter large enough to admit the end of the pencil.

ENDOCERVICITIS is known to us by the synonyms of *cervical*

catarrh and *cervical leucorrhœa*. It is a chronic inflammation of the mucous membrane, extending from the os internum to the os externum, and over the vaginal portion of the cervix uteri.

Dr. Sims (*Uterine Surgery*) gives the following reasons why *cervical leucorrhœa* is a cause of sterility:

"It is almost always of albuminous consistence, and very difficult of removal. Under the microscope it presents all the characteristics of muco-pus. Sometimes it is merely an exaggerated secretion, without any abnormal qualities. *It interferes with conception in two ways, mechanically and chemically: mechanically, by blocking up the canal of the cervix, and preventing the passage of the spermatozoa; chemically, by poisoning or killing them.* I have frequently seen conception happen while using the *nitrate of silver* for granular erosion of the os and cervix uteri. Unless there is some special reason for it, I never interdict sexual congress during the treatment of ordinary cases of cervical engorgement. When conception has taken place under these circumstances, I am satisfied that sexual intercourse must have occurred within ten or twelve hours after the use of the remedy, or at least before its eschar began to separate, which is always attended with a secretion of muco-pus, that would be fatal to the spermatozoa.

"If I were asked what, next to mere mechanical obstruction of the cervix uteri, constitutes the greatest obstacle to conception, I have no hesitation in saying that it was an abnormal secretion from the cervix. We often see the cervical mucus in such large quantities that its mere abundance will mechanically prevent the passage of the semen to the cavity of the uterus. Sir Joseph Oliffe has informed me of the case of the wife of a medical man, who had been sterile for many years, and whose cervix uteri had always presented a little mass of ropy mucus hanging from the os, that obstructed mechanically the canal. At last the doctor had the rational surgical idea to exhaust the cervix of its inspissated mucus; and sexual congress with his wife, immediately afterwards, was followed by conception.

"I am now satisfied that the cervical secretion is often poisonous to the spermatozoa, even when it would seem to be almost normal in appearance. This must depend upon some other quality than mere alkalinity, for I have often found all the spermatozoa in the cervical mucus dead, while it manifested no unusual degree of alkalinity when tested by litmus-paper; but when placed under the microscope, it showed an uncommon number of epithelial scales. This demonstrated an abnormal action in the glandular apparatus, that gave rise to this secretion, which seemed to kill the spermatozoa more by its density than by its chemical action ; for I have noticed that they lived longer in that portion of the mucus that had the fewest number of epithelial scales, and, *vice versa*, died quicker in that portion that had the most, and that, too, when litmus-paper showed no difference in the chemical character of the two."

Professor Meigs, of Philadelphia, adds his testimony to the effect of cervical leucorrhœa in preventing conception. That author observes* that some cases of sterility are caused by a plug of viscid lymph filling the canal of the cervix, so obstructing the passage as to render it apparently impossible that any spermatozoa could obtain access to the uterine cavity. "Certain it is," he says, "that some sterile women are always affected with this excessive albuminous production. Surprise has often been expressed on observing that married women, after years of sterile cohabitation, have suddenly become fruitful. In these instances, the want of fruitfulness could not depend on failure of the ovulations. May it not be that the spontaneous cure of a protracted and subacute inflammation, of the kind herein treated of, may have restored the health, and so given power to take away the woman's reproach?"

Treatment: Guernsey gives the characteristic indications for seventy-two remedies, the same as recommended in uterine catarrh.

I have found most useful the medicines which were mentioned in *endometritis*. I have also used topical applications of some of the remedies to the canal of the cervix, with Sims's applicator, or my intrauterine and cervical applicator; or placed them in apposition to the os uteri, and with the effect of hastening the curative process. I adopted this plan several years ago,

---

* On Diseases of the Uterus, p. 56.

and I am every year better satisfied with its usefulness. It has been thought, and is still believed by many of our school, that

Fig. 42.

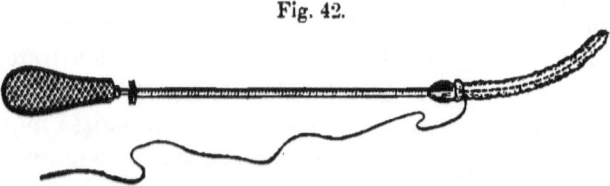

Sims's cervical applicator with cotton attached by a thread.

topical applications to the cervix, by any method, could be productive of but little benefit. But no one who has tested them thoroughly can doubt their efficacy. Remedies thus applied act by osmosis, and produce not only a local but in some cases a constitutional effect. Dr. W. H. Holcombe, one of the most logical thinkers of our school, believes in this method of applying properly chosen remedies.*

My applicator, as figured in the cut, is made of hard rubber, with a whalebone rod, ten inches long. The tube, $c\ d$, is the size of a No. 7 catheter (Eng.), which carries the rod $a\ b$, extending two and a half inches beyond its extremity, represented by

Fig. 43.

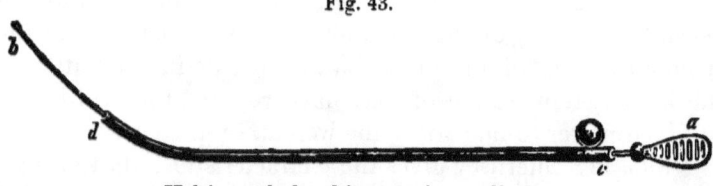

Hale's cervical and intrauterine applicator.

the space $d\ b$, the rod is flattened from $d\ b$, and terminates in a rounded blunt end. In using it the cotton is wrapped around the exposed portion of the rod, with a strong thread attached (as in Sims's applicator). This cotton is saturated with the medicinal agent selected, and if the application is to be made to the *cavity of the uterus*, the cotton is drawn back into the tube before it is introduced, when it is protruded and left in the

---

* Nature and Limitations of the Homœopathic Law.

uterine cavity, or withdrawn after applying it to the endometrium.

If the application is to be made to the *cervical canal*, the cotton is wrapped as above, medicated, and introduced up to the end of the tube. The tube is then pushed, by means of the ball or ring near the handle, until the cotton is pushed off, when both the tube and rod can be withdrawn, leaving the medicated cotton lying in the canal to be removed in a few hours.

In both applications, the cervical canal may have to be previously dilated by means of sea-tangle tents.

This instrument will be for sale by Bœricke & Tafel.

If *Sabina, Conium, Hydrastis, Senecio*, or any other remedy is indicated by the symptoms, the nature of the discharge, and the pathological state, give the remedy internally by the mouth, and apply the same to the os uteri, or even the whole length of the cervical canal.

I have had patients who could distinctly taste *argentum nitrate, iodine*, and other medicines, when thus applied to the cervix. Dr. Sims says his patients complain of the taste of *tannin* a few minutes after its application to the cervix uteri.

ABRASION, EROSION, ULCERATION.—Although these lesions of the os and cervix uteri are generally considered causes of sterility, they are not by any means invariably so.

Whitehead, Gardner, and Sims enumerate *ulceration* as a cause of sterility, but admit that it is oftener a cause of miscarriage. Nevertheless, women do often conceive, and go through the full period of pregnancy, with the os and cervix more or less the seat of one of the varieties of ulceration. It is only when the engorgement, poisonous discharge, or occlusion of the os, which sometimes attends ulceration, prevents the spermatozoa from passing into the uterine cavity, that sterility results from this condition.

According to Donne, the spermatozoa live in pus and blood. Sims says he "has frequently seen conception to happen when the cervix uteri was in a state of profuse suppuration, so that pus, *per se*, was no hindrance to this."

In the early years of homœopathy, it was a dogma of that school that no application whatever should be made to a diseased surface. Consequently, in all the old text-books of our school, no topical application to a diseased uterus was mentioned, unless with disapprobation. The immediate followers of Hahnemann failed to see that an agent may act according to the law of *similars* even when applied to the diseased surface externally. I regret that there are a few antiquated members of our school who still cling to this absurd notion. To deny that a weak, mild application of *nitric* or *chromic* acid to an ulcerated surface is not homœopathic, is to deny the truth of the homœopathic law, which is not limited, but universal in its application.

During the first ten years of my practice, I followed faithfully the dogma above referred to, and although I faithfully prescribed the carefully-selected remedies then known to our school, I never cured a case of ulceration or erosion of the os uteri. Even when I supplemented the internal medication with vaginal injections of water, pure or medicated with *Calendula* and other medicines, my success was but little better. Afterwards I had better success when I touched the diseased surface with glyceroles of *Calendula*,* *Hydrastis*, *Thuja*, etc.

Not content with this treatment, which failed in a majority of cases, I took the rational ground of applying certain agents which I knew would cause similar conditions of the surface if applied in strong solutions, or the crude substance.

I commenced with the nitrate of silver, at that time the best known and most generally used of all the eschorotics. At first I used solutions of varying strength, and with them made my first satisfactory cures. I soon learned, however, that this agent was not suited to all cases.

The sphere of action of *Argentum nit.* is confined to those cases where the cervix has a deeply congested, red, and angry appearance. The redness is so intense to be *livid*, and presents

---

* The best and only proper preparation of *Calendula off.* for topical use should be made as follows: Calendula flowers, ℥j; Water, ℥iv; place in an open-mouthed bottle and allow it to stand in a warm place six hours, then add ℥iv of glycerin, and macerate several days; strain or filter.

a velvety appearance. It bleeds readily and discharges a muco-pus, although often no discharge occurs. The menses are too frequent and profuse; there are sharp quick stitches in the cervix and vagina. The vagina is often intensely red and eroded, and an irritating, corrosive leucorrhœa is present.

In such cases, a sponge moistened with a solution of five or ten grains of the crystals to ℥j of water, applied lightly every four or five days, will often act very happily. The solution should be strong enough to coagulate the albumen on the surface (after all the mucus and pus has been carefully wiped off). But I now adopt the following methods:

(a.) I rarely use the solution of nitrate of silver, but prefer, in nearly every case, to touch the erosion, abrasion, or ulcer with the solid stick. The application should consist of light, quick, delicate touches of the blunt rounded stick. Never use the pointed stick, and never allow it to remain in contact with the surface more than a second.

Before the nitrate or any other drug is applied, the surface should be wiped dry of all secretions adherent to it.

(b.) Instead of *lint* as a dressing to the ulcerated surface, cotton saturated with pure Glycerin (diluted ½), medicated with the remedy indicated by the condition. To prepare a dressing to apply to the ulcerated surface, which has or has not been touched by the nitrate, take some fine cotton,* made into a ball, with a strong thread passing through it, immerse it in tepid water, and squeeze it gently under the water till it becomes perfectly wet; then press all the water out of it and saturate with pure Glycerin, or a medicated glycerole. To do this, lay the moistened cotton in the palm of the hand, spread it out circularly for an inch and a half in diameter, more or less, as may be needed, scooping it out in the centre; then drop half a teaspoonful of Glycerin on it, thus held, and rub it into the cotton with the point of the finger, then pour a little more Glycerin and rub it in, and so on until the cotton becomes saturated. When finished, the cotton should feel soft and

---

* There are now two elegant preparations of cotton, far better than the ordinary cotton-wool, namely, "borated cotton," and "absorbent cotton."

pulpy, should be about an inch and a half in diameter, and about half an inch thick. This dressing should be applied immediately after the application of the nitrate, and before the speculum is removed. When I use medicated Glycerin, it is usually of the proportion of *one drachm of the mother tincture to one ounce of Glycerin.* For dressing after the nitrate, the *Calendula glycerole* is best; *Hamamelis glycerole* when there is a varicose condition of the cervix with bleeding. Hydrastia mur. glycerole is often very useful. (See list of preparations on page 157.)

There are other agents, however, which I now prefer to the nitrate. In fact, I rarely use that caustic in my practice preferring in nearly all cases Chromic, Carbolic, and Nitric acids.

Chromic acid is superior to all others in granular erosions, which are very irritable, bleed readily, cause swelling of the os and cervix, bloody purulent leucorrhœa, frequent and profuse menses, sometimes as often as every fifteen or eighteen days. In all cases it should be applied strong enough to cause the diseased surface to assume a *yellow* color. A solution of five grains to ʒj of water will suffice. In severe and more obstinate cases, thirty grains to ʒj must be used. Apply with a bit of cotton closely wound on the end of a flat silver applicator. Touch lightly only the diseased surface, which has been thoroughly wiped off. After it has dried, apply the cotton tampon moistened with a Calendula or Hydrastis glycerole. Once in seven days is often enough to apply this agent.

Nitric acid is indicated when the granulations are intensely red, livid, and irritable, or where actual ulcerations exist. The ulcers are round, well defined, with sharp edges, and a lardaceous bottom. It is equally useful in the *fissured* ulcer. If the ulcers are syphilitic it is quite efficient, although it may have to be superseded by the acid nitrate of mercury.

Carbolic acid is called for when the ulceration is superficial and spreading; when the granulations are large, flat, or pointed, but pale and flabby. The discharge from the diseased surface is purulent, fetid, and often sanious. The strength of the solution should vary with the intensity of the disease. If recent, and the granulations small, a solution of ten grains of the crystals to ʒj of water will suffice. In older cases, ʒj of the acid to ʒv of water will have to be used. The best dressing to follow the application of carbolic acid is

>Kennedy's Ext. Pinus canadenis, ʒj. (colorless).
>Glycerin, ʒiij.

Apply on cotton, to be removed in twelve hours. This dressing alone will cure some mild, recent cases of erosion or abrasion. There are many other agents which are excellent applications, when applied on the cotton tampons, namely:

>Nitrate of Bismuth, grains 60, to Glycerin, ʒj.
>Tannin ʒj, to Glycerin, ʒj.
>Tinc. iodine, ʒj, to Glycerin, ʒss.
>Nitrate of sanguinaria, grains v, to water and Glycerin, of each, ʒj.
>Muriate of hydrastia, grains x, to water and Glycerin, of each, ʒss.
>Merrill's fluid hydrastia, ʒj, to Glycerin, ʒj.
>Tinc. thuja, ʒss, to Glycerin, ʒss.
>Tinc. eucalyptus, ʒss, to Glycerin, ʒss.
>Chloral hydrate, grains x, to water and Glycerin, of each, ʒss.
>Tinc. Grindelia, ʒj, to Glycerin, ʒj.

If Glycerin is inadmissible, owing to some idiosyncrasy of the patient, Cosmoline or Vaseline can be substituted with advantage for it.

Kali bichromicum (1 gr. to ʒj), in deepseated ulcerations, with a red margin and fungous appearance. Many others could be named, but the physician must carefully select the medicine which is indicated by the condition and symptoms.

(c.) When these *acids* are used, they should not be applied oftener than every *five* days. The medicated dressings should be used *once* a day. The patient can be instructed to apply them in the morning, and remove them at night on going to bed, or *vice versa*. When these dressings are used no injections are needed for cleanliness; for, as remarked by Dr. Sims, the Glycerin "seems to set up a capillary drainage by osmosis, producing a copious watery discharge, depleting the tissues with which it lies in contact, and giving them *a dry, clean, and healthy appearance*. When such a dressing is applied to a pyogenic surface on the cervix uteri for a few hours, and then removed, the sore will be as clear of pus as if it were just washed and wiped dry." Dr. Sims gives a cut illustrating an instrument, the "porte-tampon," which he says any woman can use with ease.* The tampon is placed in the open cavity, the door shut, the instrument introduced as far as possible, the tampon pushed out by the piston, and the porte-tampon withdrawn. There are now several new instruments for this purpose. A Fergusson's speculum may be used and the tampon pushed up through it by means of a stick the size of a lead-pencil.

Fig. 44.

Sim's Porte-tampon.

LEUCORRHŒA in general cannot be considered as a cause of *sterility*. Such a generalization is not in accordance with the spirit and scope of modern scientific investigations into uterine pathology. Leucorrhœa may be said to consist of three varieties:

1. A discharge from the cavity of the fundus uteri. (Endometritis.)
2. A discharge from the canal of the cervix. (Endocervicitis.)
3. A discharge from the mucous membrane of the vagina. (Vaginitis.)

The *first* variety has been considered in another place, namely, in the preceding paragraph, relating to Endometritis.

The *second* variety has been treated of above, under the head of Endocervicitis.

The *third* variety will be considered under the head of Vaginal Causes of Sterility.

DYSMENORRHŒA.—This condition, as a cause of sterility, will also be found treated of under the heads of "Stricture or Occlu-

---

* Uterine Surgery, p. 295.

sion of the Cervix," "Uterine Displacements," "Uterine Tumors," and "Inflammation of the Uterus."

Dysmenorrhœa is divided into six varieties; of these, two only are *constant* causes of sterility; namely, the *obstructive* and the *membranous*. The other four are only exceptional causes; namely, the *neuralgic, spasmodic, congestive,* and *inflammatory*.

Treatment.—Anything like a thorough consideration of the treatment of these varieties of dysmenorrhœa, would occupy too much space in this volume. A brief enumeration of the most useful remedies will have to suffice:

I. For *neuralgic* dysmenorrhœa, *Aconite, Agaricus, Belladonna, Cimicifuga, Ferrum, Gelseminum, Pulsatilla, Platinum, Senecio, Scutellaria, Xanthoxylum, Zinc, Viburnum,* etc.

II. The *congestive* requires *Aloes, Aletris, Belladonna, Cimicifuga, Borax, Sabina, Hamamelis, Lilium, Murex, Secale, Sepia, Ustilago,* etc.

III. The *spasmodic* requires *Caulophyllin, Cannabis ind., Nux vomica, Ignatia, Viburnum, Secale, Cimicifuga, Ustilago, Iodoform,** etc.

IV. The *inflammatory, Aconite, Belladonna, Gelseminum, Veratrum viride,* for acute, and a host of remedies, too numerous to mention, for chronic.

The so-called *pseudo-membranous* has been cured with *Bromine, Guaiacum, Borax,* and a few other remedies. Guernsey gives the characteristic, or key symptoms, of *ninety* medicines, for the cure of all the varieties.

AMENORRHŒA.—The pathological conditions resulting in amenorrhœa may cause sterility, but not the amenorrhœa itself.

"I do not know," says Dr. Sims, "that conception has ever occurred previously to the menstrual flow. . . . Many women conceive without menstruating, but it is always during menstrual life." There are on record cases of conception occurring after the change of life has been passed ten or fifteen years. It may sometimes occur during a temporary amenorrhœa of a few

---

* See Cincinnati Medical Advance, 1878; also Therapeutics of New Remedies.

months. (See Jackson's paper on Ovulation, at the beginning of this work.)

If amenorrhœa arises from a failure in the function of ovulation, conception cannot occur. (See "Atrophy of the Ovaries," and other ovarian disorders.)

If it results from phthisis, or serious chronic diseases, sterility is an accompaniment. If from an atonic or torpid condition of the ovaries or uterus, the resulting sterility may be cured by remedies homœopathic to the condition; such as *Sabina, Cantharis, Bromine, Aletris, Cannabis indica, Phosphorus, Helonias, Ruta, Agnus castus, Conium,* etc.

If the retention of the menses result from some mechanical cause existing in the cervix or vagina, consult the paragraph treating of stricture, occlusion, etc.

MENSTRUAL IRREGULARITIES.—It sometimes happens that certain irregularities in the quality, quantity, and time of appearance of the menses, are the only symptoms attending sterility which we are permitted to know. Cases will now and then occur, when no examination by the touch or speculum will be submitted to, and we are compelled to prescribe for the symptoms as related to us. The chief remedies for menstrual irregularities are:

For scanty menses,—*Conium, Agnus castus, Caladium, Aletris, Cocculus, Secale, Sabina, Graphites, Natrum mur., Senecio,* and *Sepia.*

For profuse menses,—*Sabina, Erigeron, China, Ipecac, Crocus, Platinum, Nux vom., Hydrastis, Trillium, Murex, Senecio, Ruta, Secale.* (See the indications for the hundred or more remedies, in Guernsey's *Diseases of Women.*)

Too frequent and profuse menses are generally controlled by *Calcarea, Senecio, Platinum, Crocus, Sabina, Ustilago, Secale, Lilium,* and *Xanthoxylum.* When this condition is unattended by any lesion of uterus the *Bromide of Ammonium,* 5 grains, three times a day, will regulate it.

For delaying menses,—*Pulsatilla, Senecio, Aletris, Natrum mur., Graphites, Helonias, Cimicifuga,* etc.

Nearly all these irregularities depend on some pathological

condition of the generative organs. It will be necessary, then, when possible, to ascertain the nature of the causes, and consult the treatment of the conditions in the preceding pages.

AREOLAR HYPERPLASIA OF THE UTERUS.*—This disease of the uterus, once known as chronic parenchymatous metritis, is one of the most intractable with which we have to deal. The great pathologist Kolb defines this condition of uterine areolar hyperplasia as "a diffuse growth of connective tissue which constitutes the so called induration, hitherto considered as a result of parenchymatous inflammation of the uterus. . . . The whole uterine connective tissue sometimes proliferates, either without accompanying increase of the muscular substance, or if this does occur, the connective tissue predominates to such an extent that the muscular tissue is comparatively of little account."

This condition has also been designated as "uterine hypertrophy," "enlargement of the cervix," and "subinvolution of the uterus" (Simpson), also "chronic engorgement of the uterus and cervix."

No morbid condition has given gynæcologists so much trouble. Allopathists and homœopathists alike find it obstinate and intractable.

I will not attempt to describe the appearance and symptoms, but refer the physician to the exhaustive and graphic description given by Thomas, *Diseases of Women*. I will, however, quote his condensation of the causes, symptoms, etc., namely:

Predisposing Causes.

A depreciation of the vital forces from any cause.

Constitutional tendency to tubercle, scrofula, or spanæmia.

Parturition, especially when repeated often and with too short intervals.

Prolonged nervous depression.

A torpid condition of the intestines and liver.

---
* This paper was presented to the Massachusetts Surgical and Gynæcological Society, at the June meeting, 1878. With some additions I have incorporated it into this work.—II.

UTERINE CAUSES.

Exciting Causes.

Overexertion after delivery.
Puerperal pelvic inflammation.
Laceration of the cervix uteri.
Displacements, endometritis, neoplasms, cardiac disease.
Abdominal tumors pressing on the vena cava.
Excessive sexual intercourse.

Symptoms.

If the cervix alone be affected there are:
Pain in the back and loins.
Pressure on bladder or rectum.
Disordered menstruation.
Difficulty of locomotion.
Nervous disorder.
Pain on sexual intercourse.
Dyspepsia, headache, languor.
Leucorrhœa.
If the body of the uterus is affected there are:
A dull, heavy, dragging pain through the pelvis, much increased by locomotion.
Pain on defecation and coition.
Dull pain beginning several days before menstruation, and lasting through that process.
Pain in the mammæ, before and during menstruation.
Darkening of the areolæ of the breasts.
Nausea and vomiting.
Pressure on the rectum, with tenesmus and hæmorrhoids.
Pressure on the bladder, with vesical tenesmus.
Sterility.

Physical Signs.

If *cervical hyperplasia* exists, there will be:
Sinking of the uterus in the pelvis.
Large, swollen, and sensitive cervix.
Normal uterine axis changed, the fundus tipping forward or backward.

If *corporeal hyperplasia* exists, there will be:
Uterus enlarged, heavy, and sensitive.
The uterine cavity enlarged.
Sensitiveness of the internal uterine surface.
Uterine hyperplasia can only be confounded with:
Pregnancy.
Neoplasms, fibrous growths in uterine wall.
Periuterine inflammations.

Treatment.—Putting aside the treatment of the various causes and complications connected with this condition, and confining ourselves to the special treatment of the *areolar hyperplasia*, what medicines are indicated in that pathological state? We must first know what the pathological condition really is which obtains in this disease. It is divided into two stages, and although the same remedies may be indicated in both, the dose and method of administration will differ. The *first stage* is characterized thus: The hypertrophied areolar tissue is congested, containing absolutely more blood than normal, and the whole of the affected part, neck, body, or entire uterus, is greatly increased in size and weight, and the cavity increased in size. This stage may last months, or even years, but as time passes the *second stage* supervenes, and an opposite state of things is set up. The large, red, soft and engorged uterus decreases in size and becomes small, contracted, hard, white, and anæmic, and the cavity *decreased* in size.

Kolb thus describes advanced cases: "The parenchyma on section appears whiter, or of a whitish-red color, deficient in bloodvessels from compression of the capillaries, by the contraction of the newly-formed connective tissue, or from partial destruction or obliteration of the vessels during the growth of tissue. The firmness of the uterine substance is also increased, simulating the hardness of cartilage and creaking under the knife." This constitutes a true *sclerosis* of the uterus.

Now, in order to select the appropriate remedies, the practitioner must possess an extensive and thorough knowledge of Materia Medica, such as only studious members of our school can possess. He must not only have a knowledge of the *symp-*

toms of drugs, but the *pathological* conditions they are capable of causing in the uterus or analogous organs. No medicine can be useful in areolar hyperplasia, unless it causes the *first* stage as well as the *second*, and the first stage (of congestion, etc.) must have preceded the second. In other words, *the history and order of sequence of the medicinal disease must be similar to that of the natural one.*

I believe there are cases of this disease when the second stage I have described may have been the only recognizable stage, as in those cases of "morbid excess in the involution of the uterus after parturition," described by Simpson. But these cases are rare, and require peculiar treatment.

The medicines which are useful in this disease are those which are capable of causing both stages of the pathological condition. They must first cause

Congestion of the uterus, or
Inflammation "   "
Enlargement   "   "
Œdema   "   "
Sensitiveness "   "
Increased weight of the uterus.
    "   capacity "   "
Painfulness of the uterus.
Too frequent and profuse (or scanty) menses.

With the various concomitant or reflex sufferings which accompany this stage.

They must also cause the following conditions:

Anæmia of the uterus.
Œdema   "   "
Sclerosis "   "
Hard, small, and contracted uterus.
Scanty and delaying menses.

Together with the various pains and sufferings attendant on such a condition of the uterus.

Thomas and other prominent authors do not appear to lay down any treatment for this second stage, either because they consider it not amenable to treatment, or because there is no necessity for any treatment. I propose to give such treatment for the second stage as the materials of our Materia Medica permit, because it causes one condition, namely, *sterility*, which we are often called upon to treat, and with such indifferent success.

The medicines generally indicated for the *first* stage of the disease are the following (those italicized are the most important):

Apis.	Cauloph.	Podoph.
Asarum.	*Cimicifuga.*	*Platinum.*
Aurum.	Calc. carb.	Ruta.
Ammonium carb.	*Erigeron.*	Sanguinaria.
Argentum nit.	Ferrum.	*Sabina.*
Arsenicum.	*Hamamelis.*	Sepia.
Æsculus.	*Helonias.*	*Secale.*
Belladonna.	Lachesis.	Trillium.
Borax.	*Lilium.*	Terebinth.
Cantharis.	Merc. cor.	*Tanacetum.*
Cannabis ind.	Murex.	*Thuja.*
Cocculus.	Nux vom.	Senecio.
Crocus.	Nux mosch.	*Ustilago.*
Cactus.	*Pulsatilla.*	Sulphur.

All these medicines, by their *primary* action, cause the following conditions of the uterus:

Irritation.	Inflammation.
Congestion.	Enlargements.

NOTE.—Prof. R. Ludlam (U. S. Medical Investigator, Nov. 1877) recommends *Tartar Emetic* 3x in "corporeal cervicitis, with concentric hypertrophy due to effusion of serum into its tissues." He says he has used it successfully for ten years, yet he does not mention it in his "Diseases of Women." He also asserts it to be one of the best internal remedies for "catarrhal inflammation of the glandular portion of the cervix." The conditions for which he recommends it are not exactly "areolar hyperplasia," although they may be precedent or concomitant conditions. The first might be called "œdema of the cervix," for which, on page 186 of this work, I have advised Apis, Ars. iod., Polymnia, etc. From some recent experience, I am inclined to believe that Antimonium *iod.* or Antimonium *ars.* will prove more useful than Antimonium *tart.*, especially in chronic cervicitis with œdematous enlargement.

Attended by:
   Sensitiveness and tenderness of the uterus.
   Increased weight   "   "
   "   capacity   "   "
   Painfulness   "   "
   Too frequent menses (with menorrhagia), or,
   Too scanty menses.
   Dysmenorrhœa.
With all the reflex symptoms belonging to such conditions.

These remedies should be carefully selected by the symptoms, and also by the "genius" of the medicine.

They should be prescribed in the attenuations from the 6th to the 30th, and not too often repeated.

Many of them may be applied to the uterus on cotton tampons, by diluting them to the strength of 1x or 2x, using glycerin as a vehicle.

There are a few exceptions to this rule, namely, in cases of *subinvolution* after parturition or miscarriage, when, if the carefully selected remedy does not bring about a rapid change, we should resort to those medicines which act in material doses in such a way as to contract the bloodvessels of the uterus, and thus cut off the nutritive supply, and at the same time contract the muscular fibres to decrease the size of the engorged and soft parenchyma. There are but few medicines capable of such action, namely: *Bromide of potassa, ammonium, lime, soda* etc., *Caulophyllin, Secale, Ustilago,* and perhaps *Gossypium* and *Viscum alb.* These should be given in the following manner:

Bromide of potassa, etc., 5 to 10 grains every six hours.
Caulophyllin, . . . $\frac{1}{8}$ to 1 grain "   "
Secale, ... 5 to 30 drops of a good extract every six or twelve hours, or 10 grains of the 1x trit. of *ergotin* every four hours.
Ustilago, . 5 to 10 grains of the 1x trit. every two or four hours.
Gossypium or Viscum album, } . . 5 or 10 drops every two or four hours.

In the treatment of the *first* stage there are certain very important auxiliary measures which must be prescribed, for without them medicines can do little.

1. In severe cases *rest* in the recumbent posture for a few hours each day, with a few hours' walking or riding in the pure open air. The removal of all pressure from the abdomen from tight corsets, heavy skirts, etc.

2. The use of plain, non-stimulating food and beverages, and an open state of the bowels.

3. Nearly complete abstinence from sexual intercourse; at most, a few.times midway between the menstrual periods.

4. *Enemas of hot water* of a temperature of 100° to 110° F., and in quantity not less than a quart, or more than three or four gallons. This should be thrown in a steady stream against the cervix uteri, by means of an ordinary pump syringe, or a fountain syringe; the latter is the best, as no labor of the hand is required, and the reservoir can be as large as required. It can be used once a day—at night—or in obstinate cases, twice daily. Those who have never tried this method will be surprised at the change in size and color—a dimunition of both—which will take place after the use of the hot water for a week or two. It is now the standard and popular practice of all the best gynaecologists.

5. *Local depletion* by means of *leeches*, or the *scarificator*. In the early years of my practice, I closely adhered to the teachings of the leaders of our school at that time (1850 to 1865). But my success in treating uterine disease was neither satisfactory to myself or my patients. I then resolved to adopt the modern teachings of those eminent gynaecologists, who, while they discarded the general depletions, painful caustics, and other horrible measures of the allopathic schools, imitated nature's processes as far as possible. Nature causes a congested uterus to bleed to relieve the engorgement, or contracts, by means of vaso-motor nerves, the dilated bloodvessels. The application of a single *leech* to the os uteri relieves the local engorgement, and allows our remedies to act better than they could possibly do before. *Scarification* of the os and cervix does the same. But I prefer to either the *spear*. This little

instrument, first introduced to notice by Dr. Buttles, bears his name. Its sharp, delicate point is thrust into the enlarged and congested cervix, about the $\frac{1}{16}$ of an inch, and given a single turn before it is removed. Six or eight punctures will cause a flow of from half to one ounce of blood, according to the severity of the case. There are certain indications for the use of local depletion, and they should be closely followed, namely:

Fig. 45.

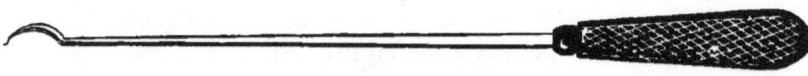

Chapman's scarificator.

When the woman complains of sensations of *fulness*, *weight*, *soreness*, and *aching* in the uterus, it should be used. Now these sensations generally occur every week or ten days, during the intermenstrual periods, and denote recurring congestions. They always precede the appearance of the menses, which may occur every two or three weeks.

It has been my practice for several years to use the *spear* whenever these symptoms are complained of, and the relief given to the patient is always gratifying.

Fig. 46.

Buttle's spear-scarificator.

By this means I have regulated the appearance of the menses, that for years had occurred *too often*. It prevents profuse menorrhagia, and causes the too scanty and painful menses to be natural in quantity. The uterus in this disease may be so congested as to bleed too much (menorrhagia), or so engorged as to bleed too little, or not at all (suppression of the menses).

In either case if we relieve the local congestion, we remove the cause of the abnormal menstrual flow. This cannot be done by medicines alone, unless absolute rest be adopted, and then it requires a long time. It must be noted that I do not refer to other conditions of the uterus, or ulceration, atony, etc., but only to the *first* stage of areolar hyperplasia.

As an illustration of the value of *puncting* the cervix in certain cases of sterility, I quote from a paper by Dr. A. R. Jackson, a prominent writer on that subject.

"In the summer of 1876, I was consulted by a lady thirty years of age. Menstruation had always been rather scanty, the periods lasting two or three days. She had been married nine years and had never been pregnant so far as she knew, although during the first year of her married life she had "missed" one period, and this was followed by nausea, headache, and some other evidences of nervous derangement. However, without any treatment the next period came at its regular time, marked only by a somewhat unusually profuse flow. Within the past three years she had grown very stout and plethoric, and the catamenia, although quite regular, had become steadily more scanty, until, finally, the use of a single napkin was sufficient. She attributed her stoutness to the diminished quantity of the discharge, and it was for the purpose of having the latter increased that she sought advice.

In reply to a question, she expressed a strong desire for children, but stated that she had long since abandoned all hope of ever having them.

Examination revealed nothing abnormal.

I advised her to take very active exercise, to use a restricted diet, to take mild saline laxative medicines, and recommended the local abstraction of blood.

On the day prior to the expected appearance of the menses, I punctured the uterine cervix deeply, so as to remove two or three tablespoonfuls of blood. The regular discharge appeared within twelve hours, and the quantity passed was more than double that of the period immediately preceding. This operation was repeated just before each menstruation subsequently.

The treatment was continued four months, at the end of which time the weight of the patient was reduced fifteen pounds, and the catamenia had so increased that, at the fourth period, she used five napkins. The fifth failed to appear. Conception had occurred, and was followed in due time by the birth of a healthy girl."

Local Depletion as a Cure for Sterility.—But the most valuable and important result which often follows the use of local depletion, is the restoration of the ability to conceive.

Long before I adopted its use, I was acquainted with numerous cases, several of them former patients of mine, who after a few applications of leeches, became pregnant, much to their surprise. Some of these patients had never before been pregnant; others had not borne children for many years. One case in particular impressed me profoundly, a lady, married fifteen years, who had never been pregnant. She had been treated by eminent men of both schools. She was under my care nearly a year; there was no abnormal appearance visible, except enlargement and chronic congestion. The menses were delaying, very *scanty*, and very painful. After she left my care she was advised by an English nurse to apply a leech to the uterus about the time the menses ought to appear. She became pregnant after the second application. Three years afterwards she suffered from a return of her old troubles; leeches were applied, followed by another pregnancy; and this occurred the third time! This patient informed other sterile women of this fact, and some of them tried it with success. I have the records of several cases where the use of the *spear-scarificator* was soon followed by pregnancy in women who had been sterile for several years.

Another method of local depletion which should not be neglected, is by means of tampons of medicated or pure Glycerin. I have already given the uses of these in another place; also the method of their application. It remains only to point out the condition requiring them.

In many cases the cervix presents a swollen, puffy, and pale appearance. It looks *œdematous*, and is in fact œdematous. It is engorged with serum instead of blood. A puncture with the *spear* brings but little blood, but a watery exudation. I have seen cases in which the cervix actually *pitted* on continued pressure. The medicines indicated in this condition are *Apis, Aquaphobin, Arsenicum, Cantharis, Hamamelis, Iodine, Lachesis, Naja, Murex, Merc. cor., Phytolacca, Sepia,* and *Sulphur.* The action of Glycerin when applied continuously to such a cervix is to cause profuse drainage, or exosmosis of water from the œdematous tissues. It should be applied undiluted, night and morning. The addition of one-tenth or $\frac{1}{100}$ part of Iodine, Iodide of potassa, Hamamelis, Cantharis, Merc. cor., or Phytolacca, also *Polymnia*,* greatly increases the value of the Glycerin in removing the œdematous enlargement. If the whole body of the uterus is believed to partake of this dropsical condition, a small medicated tampon may be placed in the uterine cavity by means of a proper applicator. I have several times seen excellent results follow their use, in addition to the external tampon.

* *Polymnia uvedalia* (Bearsfoot) belongs to the genus *Compositæ*, of the tribe *Helianthew*. It is an erect herb, roughish, hairy, stout, 4 to 10 feet high, leaves broadly ovate, angled and toothed, nearly sessile, the lower palmately lobed; abruptly narrowed into a winged petiole, outer involucral scales very large; rays 10 to 15, linear oblong, much larger than the inner scales of the involucre; flower *yellow*. Grows in rich soil west of New York to Illinois, and southward. The flower and whole plant exhale a strong odor; they look like a small sunflower.

This remedy was introduced into eclectic practice by Dr. Pruitt, of Missouri (?), who found it useful for enlargement of the spleen. He recommended that it be applied externally in the form of an ointment and given internally. It was found so efficacious in this disease, that it was tried in other enlargements, of the joints, glands, etc., and finally in *enlargements of the uterus.* I have used it in two cases with success—in uterine hypertrophy (areolar hyperplasia) in the first stages, before condensation and contraction set in. The dose is 10 drops of the tincture or 1x, three times a day, or oftener, and its use continued for months; at the same time apply a glycerole (1 part of the tincture and 2 parts glycerin) in cotton tampons every night except during the menses. In my two cases the uterus decreased in size fully one-third in two months.

Dr. Scudder* gives the following as his experience, with an estimate of the power of this new remedy.

... "Let me again say that Uvedalia is the straight remedy for those engorgements of tissue depending upon an enfeebled circulation. The sensation given to the fingers is a want of elasticity and tonicity, sodden, doughy, atonic.

"I use the Uvedalia ointment freely, and think of rubbing away a hypertrophied uterus, or an enlarged joint, as much as I would an enlarged spleen or ague-cake. Recently I had occasion to use it in chronic ovaritis and metritis, the cavity of the uterus measuring four inches. The organ was reduced to normal size in four weeks.

"Internally it has not gotten to be such a favorite, but this is because I have not tried it so often. I am satisfied that in many cases of chronic disease it will be found a most valuable remedy, as it very certainly is in chronic ague with enlarged spleen.

"I am sure those who have used it will feel much obliged to Dr. Pruitt for his efforts to bring it to the notice of the profession.

* Eclectic Medical Journal, February, 1878.

Dr. Thomas (*Diseases of Women*), after giving all the various methods of treatment, general and local, thus sums up by candidly giving his own experience and its results:

"The best local alterative is the compound Tincture of Iodine, which by means of a brush of pig's bristles should be carried up to the os internum, or even to the fundus should endometritis exist, and over the cervix; then waiting for a complete drying, this process should be repeated. After these applications a wad of cotton, to which a string has been attached in such a way as to leave its surface flat, should be saturated with glycerin and laid against the ·cervix. This acts as a local hydragogue and disgorges the tissues. These local applications should be repeated once a week, but others should be made oftener by the patient herself, by means of vaginal injections, by which the drug just mentioned may be brought in contact with the cervix.

"Mild and lacking in vigor as this course may appear, let any one test it side by side with the plan of using the acid nitrate of mercury, potassa fusa, potassa cum calc., and the actual cautery, etc., ... and unless his experience greatly differs from mine, he will feel that in the former he has reached a resting-place for his faith in the treatment of the most important of all the forms of uterine disease. He will see proof daily spring up before him that his capacity for benefiting his patients has greatly increased, while his liability to injure them has greatly diminished."

Treatment of the Second Stage.—The second stage of areolar hyperplasia is the stage wherein sterility is sure to occur, even if it did not occur during the first.

But it is possible, by the use of carefully selected remedies and appropriate auxilliary measures, to restore the organ to a condition in which conception is possible.

The same medicines mentioned as useful in the first stage, are also useful in the second, but they will not be successful in arousing the torpid energies of the uterus unless they are prescribed in the lowest attenuations, or in material doses. In proof of this may be cited the want of success which has attended their use in such cases when given in high potencies, and the positive success which has followed their use when given in appreciable doses, by members of our school, and of other schools of practice. Thus Cantharis, Cannabis indica, Moschus, Sabina, Phosphorus, Cimicifuga, Sanguinaria, and others have been successfully used in quite large doses, and have cured sterility due to atony, atrophy, and paretic conditions of the uterus and ovaries.

They cure, by increasing the flow of blood to the shrunken and poorly nourished organs of generation, and by imparting to them a normal supply of nervous energy. It is generally necessary to continue their use a considerable period of time, and aid their influence by the action of electricity, appropriate diet, change of climate, etc.

There is another class of medicines, however, which are capable of bringing on a condition similar to the *second* stage of areolar hyperplasia, or sclerosis, by their continued *primary* action. They are:

Arnica.	Lachesis.
Aletris.	*Lapis alb.*
Aurum mur.	Lycopodium.
Ammon mur.	Magnesia mur.
Argentum mur.	Manganum mur.
Arsenicum iod.	*Merc. iod.*
Baryta mur.	Morphinum.
Bromine.	*Natrum mur.*
Calc. mur.	Nux vom.
Calc. iod.	Plumbum iod.
Cuprum.	*Phytolacca.*
Carbo veg.	*Picric acid.*
Chimaphila.	Polymnia uvedalia.
Ferrum iod.	*Platinum mur.*
Iodine.	Palladium.
Kali brom.	Silica.
Kali iod.	*Sepia.*
Kali carb.	Sulphur.

These remedies should be prescribed in the medium attenuations and continued a long time.

The *muriates* correspond to a condition wherein the sclerosis is the predominant condition. The uterus is contracted, hard, and sometimes atrophied.

The *iodides* are useful in a similar condition, but the atrophy and hardness are not as well marked, the tissues are softer and the sclerosis does not predominate.

Arnica, when the original cause was of a traumatic nature.

Arsenicum, when there is threatened tissue degeneration.

Cuprum and Ferrum in deficiency of red blood-globules, with loss of energy in the spinal trophic nerves (also Nux, Ignatia and Phosphorus).

Lycopodium, Kali brom., and *carb., Natrum mur.*, and Plumbum, when the menses have grown gradually less in quantity, until they have nearly ceased.

Phytolacca, Chimaphila, Conium, and Iodine when there is atrophy of the mammæ.

In this stage the menses are usually very scanty and infrequent. The uterus is nearly in the same condition as after the "change of life." The ovaries may still carry on the manufacture and extrusion of ovules. The ovules may even become impregnated, but owing to the paretic condition of the uterus, they do not make a lodgment therein. The object of treatment, then, is to stimulate that organ to healthy nutrition. Should this stage appear in unmarried women, marriage, by giving the uterus its natural physiological stimulus, will often remove the condition. It is possible that the careful but persistent use of Ulmus fulva bougies, with tampons of cotton medicated with Cantharis 3d, Sumbul 1^x, or Phosphorus 2^x, or Bromine 3^x, would impart the necessary physiological life.

Electricity.—It is in this condition that the use of the stimulating Faradic current ought to be productive of great and lasting good. It should be applied by one who thoroughly understands the *modus operandi*, and applied directly to the cavity of the uterus. In these cases, also, the intrauterine galvanic pessary, once recommended by Prof. Simpson and lately by Dr. Thomas, may be found very useful.

Nutritive Treatment.—The most careful selection and administration of internal remedies, as well as the most judicious use of topical applications, will sometimes fail to cure cases of *chronic endometritis, endocervicitis,* or *chronic metritis* (areolar hyperplasia), because of the general constitutional debility or acquired fault of nutrition, assimilation, etc. Homœopathic physicians are more apt to neglect the nutritive and restorative (tonic) treatment, than are the members of the opposite schools. The senseless "diet rules" which, until lately, were followed by the stricter homœopathists, had much to do in inducing this neglect. Many suppose to this day that if the perfect medicinal *similimum* is found—*i. e.,* one whose symptoms correspond closely to those of the disease or the patient—nothing else was necessary to bring about a cure. This is not, however, the case. It is as much the duty of the physician to prescribe the exercise, the diet, and the bathing and other hygienic agents, as it is to prescribe medicines, or make use of surgical means.

These considerations impel me to make the following observations:

The general condition of the patient should be examined and looked after. If the woman is obese, lymphatic, with superabundance of fatty deposit, we should prohibit all starchy foods, fats, vegetables, etc., and direct her to adopt nearly the diet laid down on page 153, by Dr. Cutter, to prevent uterine fibroids. Under this regimen the patient will grow stronger, and the disappearance of obesity will allow her muscular system to become more developed by active exercise. The habitual use of mineral waters—Kissingen, Vichy, Seltzer, Hawthorn, or Sheboygan, and other foreign and native springs —will greatly facilitate the cure.

If the patient is thin, emaciated, debilitated, and the victim of some chronic cachexia, with all the functions of vegetative life in an atonic condition, with defective nutrition and deficient vitality in general, we should add to our specific remedial treatment the injunction to eat the most nutritious and easily digested and assimilated foods. Advise chalybeate waters, seabathing, exercise in the open air, the removal from a malarious climate, or from unhealthy houses. Order the use of pure wines, porter, ale, or, better than all, that purest of all stimulants and beverages, *Kumyss.*

This Kumyss or *Koumiss* is a veritable "wine of milk," and contains in itself all that is necessary to build up an anæmic, atonic, and vitiated constitution. It restores digestion, favors the assimilation of other foods, imparts phosphates and other valuable constituents to the system, and is at the same time a food, a beverage, and a tonic.

Kumyss (*Vinum lactis, Lac equinum fermentatum*) is prepared by the fermentation of mare's milk. The fermentation of milk-sugar produces alcohol, carbonic acid, and lactic acid, and to these products the action of Kumyss seems to be chiefly due.

It has now been well-established that the milk of the cow is equal, if not superior to mare's in the production of Kumyss. In fact, the milk of all animals is convertible into this beverage by one and the same method. Kumyss is now manufactured in Paris, London, New York, Cleveland Chicago, and I believe, St. Louis. In order to give the reader some idea of its constituents, I quote the chemical analysis as given by the best authorities.

New or fresh Kumyss, when a few days old, has a sweetish, acidulous taste and a sparkling effervescence, and looks like rich milk. It contains alcohol, 1.65 per cent.; fat, 2.05; milk-sugar, 2.20; lactic acid, 1.15; solids, 6.80; casein, 1.12; carbonic acid, 0.785; salts, 0.28.

Old Kumyss is not as white, has a stronger acid taste, is very sparkling and exhilarating. It contains alcohol, 3.23 per cent.; fat, 1.01; sugar, 0.00; lactic acid, 2.93; casein and salts, 1.21; carbonic acid, 1.86; solids, 5.04. All the sugar is now converted into alcohol. It is interesting to compare this with the analysis of wines and beer.

Claret wines contain from 7 to 12 per cent. of alcohol; champagne contains from 10 to 13 per cent.; cider, 5 to 9 per cent.; ale and porter, 4 to 7 per cent.; small beer, 1.28. Kumyss stands midway between small beer and the weakest ales or lager beer.

Patients can drink with advantage from one to three bottles daily. I have had patients so debilitated from chronic exhaustive uterine diseases, and with such impaired digestion and complete loss of all appetite that inanition seemed imminent, rapidly restored to health by the persistent use of Kumyss alone for weeks or months, or until they had sufficient appetite and digestion to subsist on ordinary foods.*

Next in value is the *extract of malt*, now so much used in this country. The liquid or granulated malt extract is combined with other nutrient or medicinal agents. Among the most useful I esteem the following:

    Extract of malt with Pepsin.
    "   "   Hypophosphites.
    "   "   Iron.
    "   "   Cod-liver oil.

The experience and judgment of the physician will enable him to advise the preparation most suitable to the particular condition of each patient. These preparations will not interfere with the special, specific medicinal treatment which may be instituted.

In certain *strumous* or *scrofulous* patients the use of *Cod-liver oil* is almost indispensable. It will enable us to restore to health patients who would otherwise be absolutely incurable. As stated in my *Therapeutics of New Remedies*, besides the fatty food, Cod-liver oil contains many remedial agents in minute quantities, which our school values very highly, namely: Iodine, Bromine, Calc., etc. It need not be given in the large, nauseous doses advised by the opposite school, but will cure in very small doses, 15 to 60 drops three times a day, if used for a considerable time.

When the patient, whether anæmic or not, has a feeble, or irregular and weak circulation, the use of Digitalis or Digitalin, combined or alternated with the specific tonic or nutritive remedy, will greatly increase its restorative power. Digitalis increases the force and vigor of the heart's action, and thus aids the curative force. I recommend Digitalis 1^x, five drops, or Digitalin 2^x, two grains, three times a day.

---

* Mr. A. Arend, a chemist, of Chicago, is the originator of Kumyss in this country. After many years of experiment, he has discovered a method of making it from cow's milk, so that it nearly resembles the original Tartar beverage made from mare's milk.

ABNORMAL SHAPES OF THE OS AND CERVIX UTERI.—We often see cases of sterility when there is no symptom of disease, so far as physical suffering is concerned. We may even find the uterus of proper size, in normal position, and with a straight cervical canal, but the cervix may be of an unnatural shape. There are several abnormalities mentioned by Sims, namely:

(1.) Normal shape, with "pinhole os."
(2.) Conical cervix, with or without "pinhole os."
(3.) Crescentic-shaped os.
(4.) Too close apposition of the lips of the os.
(5.) Destruction of one lip of the os.
(6.) Overlapping os; curving of the os, etc.

These modifications of the shape, size, and relations of the os tincæ may all become causes of sterility by preventing the entrance of the spermatozoa, and require surgical operations for their removal. Dilating with tents, bougies, etc., do no good; the knife is the only trustworthy remedy.

(1.) *A pinhole os* can be readily changed to an open one of normal size by the bistoury cutting slightly each way.

(2.) *A conical cervix* may be treated in the same manner, but it may sometimes be necessary to amputate the terminal portion, or make deep bilateral incisions.

(3.) *In crescentic-shaped os*, Sims cuts out a triangular portion, removing the offending lip altogether.

(4.) *Closed os* from apposition requires a bilateral incision of the circular fibres of the cervix, with such after treatment as will prevent their closure.

(5.) *In destruction of one lip* there is generally a dense fibrous condition of the cervix. This requires the same operation as for constriction of the cervix, the bilateral incision as far upward as necessary, and its subsequent dilatation.

(6.) *Overlapping cervix* requires the cutting off of the overlapping lip, which straightens the canal and opens the door for the entrance of the spermatozoa.

I have treated several cases of Nos. 1, 2, 4, and 6, and in all but a few cases the slight operations performed resulted in curing a sterility of several years. A portion of these cases were congenital, others acquired, or were caused by severe local treatment.

One case, similar to the case described by Sims, page 181, *Uterine Surgery*, where the os closed by a fibrous band, and the menses and an intrauterine catarrh of a sanious irritating character escaped through *two* small orifices about half an inch apart. This band was cut from one orifice to the other, the cervical canal dilated by means of a sea-tangle tent, and the woman, who had not been pregnant for six years, and had suffered intense agony from dysmenorrhœa, became pregnant before the next menstrual period.

I might add another variety to the abnormal forms of the cervix mentioned by Sims. It might be called the "smashed hat" cervix. It looks like a silk hat that had been sat upon. At the vaginal junction it is of the normal shape and size, but from this point to the os it enlarges, the end of the cervix is flat, the edge of the circumference sharp and high, like the base of a cone, while the cervix is corrugated. In the three instances I have observed the os was not larger than a pin's head, and the women were sterile.

The best instrument for these operations on the os for the purpose of enlarging the orifice, is that known as Skeene's Hysterotome. When closed the blade is shut against a probe, like

Fig. 47.

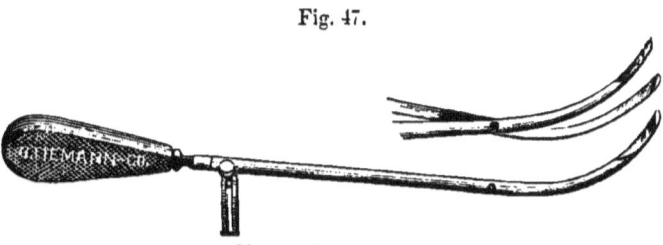

Skeene's hysterotome.

a sound, and is thus introduced into the cavity of the cervix. The blade is then opened as much as is considered necessary, and rapidly withdrawn. It is again introduced, and the other side incised in the same manner. The *outer* os should be made at least one-fourth of an inch wide. The cut surfaces should be touched with *persulphate of iron*, and a cotton tampon wet with Calendula-glycerole, be placed against it. This instrument may be used in place of White's Hysterotome, but it is not quite as safe or exact.

Another very useful and safe instrument for incising the os and cervix, is Peaslee's Uterotome. The probe-point is introduced up to the os internum, and the lancet-shaped knife is pushed up, incising both sides of the os and cervix equally.

In a large proportion of cases of *deformed cervix* with *contracted os*, I have found a *tortuous* canal. This, besides being an additional obstacle to the entrance of the semen, is an obstacle to the ready outflow of menstrual blood, and thus a cause of dys-

Fig. 48.

Peaslee's Uterotome.

menorrhœa. This condition is readily removed (after enlarging the os) by the use of the *laminaria tupelo, compith*, or *ulmusfulva* * tent, worn a few hours at a time, and placed every three or four days during the ten days previous to the menses. The patient should not walk or ride during the time the tent is worn. If she does she is liable to bring on hæmorrhage or inflammation. †

---

* Tupelo-Dilators or Tents.—(Dr. Herm. Hager.) The wood of the water-tupelo, *Nyssa aquatica* L. and *Nyssa biflora* Mich., growing along rivers and in swamps of North and South Carolina, is very light, yellowish-white, and has the property of absorbing much water in a short time, so as to increase in bulk several times. Of late the wood has been used to make tents, in the same manner and for the same purposes as those made from the laminaria. A tent of 4.75 cm. (=1¼ inches) in length, and 0.8 cm. (¾ inch) in thickness swelled up in water inside of half an hour to a length of 5 cm. (2 inches) and a thickness of 1.6 cm.(⅝ inch). The absorbed water amounts to about 5 times the weight of the *dry wood.—Pharm. Centralh.*, 1879, No. 6.

Experiments made with these tents in Switzerland have shown them to be much superior to sponge and sea-tangle tents, both of which are likely to be superseded by it.—*Schweiz. Wochensch. f. Pharm.*, 1879. No. 7.

† In even the slightest operations on the generative organs of women, especially on the uterus, the *temperature* of the body should be carefully watched. The use of the clinical thermometer is now considered indispensable in all surgical procedures, and its use should never be neglected. If the temperature in the axilla or under the tongue is above 98.6° or 99°, *no operation should be performed* until the temperature becomes normal. If after any operation the temperature rises above normal, give immediately, *Veratrum viride*, 1 or $\frac{1}{10}$ of a drop every hour or two. *No other remedy so surely prevents inflammation.*

Conoidal Cervix.—One of the simplest methods of treating unimpregnated os uteri with *conoidal cervix* and pinhole os is that recently proposed by Dr. E. L. Drake, of Fayetteville, Tenn.* It consists in placing some foreign body in the vagina so that the os uteri will rest upon or against it. A soft rubber or glass pessary, a cotton ball covered with rubber "dam," or a small, soft sponge, similarly covered, will have the effect, says Dr. Drake, of "bringing about sufficient dilatation of the os to admit the tip of the finger. Further, when the uterus, from overweight or laxity of support, presses against the perinæum, or from other displacement the os is fretted against the vaginal wall for some time, characteristic changes are set up, namely, a hypersecretion of mucus, a softening of the tissues, and a considerable degree of dilatation. Thus nature in these cases overcomes a stenosed cervix and contracted os with considerable certainty, and thereby promotes a free menstrual discharge, unless too much obstruction exists from a flexure at the internal os. The pressure of any foreign body against the os will stimulate the menstrual flux, and so become available in the treatment of amenorrhœa, which is often only a form of dysmenor-

Fig. 49.

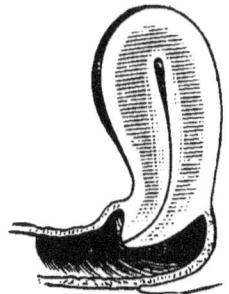
Conical cervix (after Sims).

rhœa." Dr. Drake recommends this simple method in place of rapid dilatation, incision, or amputation. The pessary should be worn for two weeks in the intermenstrual period. He claims that it will also dilate the inner os, for he has observed that the

---

* American Journal of Obstetrics, April, 1878, p. 366.

sound passes more readily to the fundus after the use of the pessary. He prefers the concavo-convex rubber or glass instruments.

The following case reported by Dr. Drake appears to substantiate his claim: "It was a typical case of conoidal cervix and pinhole os. The patient's sufferings were intense, lasting nearly through the intermenstrual period, and leaving her in a wretched condition to withstand the succeeding ones. The sympathetic disturbance was great, manifested in sympathetic cramps of the stomach and torturing headache, which were barely held in abeyance with excessive doses of Chloral and Morphine. She had been a sufferer for years, dating from an abortion. I used with her a small glass concavo-convex pessary, and have had no occasion to visit her for months; whereas before, my visits were required regularly at her periods. The last time I examined her the os would admit the tip of the finger."

Dr. Mundé, in commenting on this method, quotes a curious case in point, related by Dr. Thomas, in his *Diseases of Women*, where a globe pessary had gradually dilated the os and cervical canal, and migrated *into* the uterine cavity, where it remained for some time, causing obscure uterine symptoms, until it was detected and removed by Dr. Sayre.

My own observations corroborate Dr. Drake. I have known the constant wearing of the cotton ball, changing it every twelve or twenty-four hours, moistened with dilute glycerin, to flatten out a conoidal cervix and enlarge the os, and increase decidedly the usual scanty menses. I once placed in the canal of the cervix a Chamber's intrauterine stem pessary, for the purpose of removing a flexion of the cervix. The os was very small, the menses scanty, but after wearing the stem, the menses increased, and the os dilated to a normal size.

The following case, copied from a paper on *Sterility*, read before the Amer. Gynæcological Society, by A. Reeves Jackson, M.D., is an excellent illustration of the success which will attend the removal of a small obstacle. It aptly comes under the variety of abnormalities of the os, which I have named above as "pin-hole os." Dr. Jackson writes:

"The following case may serve to illustrate the occasional efficacy of perseverance in treatment, and also to show that comparatively simple means will sometimes succeed where elaborate ones have failed: —

"Mrs. H., aged thirty-three years, a well-formed woman of medium height, sterile, had been married eleven years. Menstruation commenced at fifteen and had always been regular, painless, and of proper amount; no leucorrhœa; general health perfect. She was sexually insensitive, and to this circumstance both husband and wife attributed the barrenness. At the end of the second year after marriage a physician was consulted, and for nearly a year she was treated by the administration of aphrodisiac medicines without result. During the succeeding three years she was almost constantly under the care of some physician of some sort, and drugs, baths, electricity, pessaries, etc., were all tried, in the vain hope that her condition might be changed. Finally, she came to me. She appeared to be perfectly well, was remarkably gay and cheerful in manner, and expressed a strong faith that something could yet be done for her.

"On examination I found all the genital organs normal as to development, position, and state, with the single exception that the os uteri was extremely small; indeed, through the speculum its site was indicated only by a reddish spot no larger than a pin's head. On discovering this I questioned her more particularly in regard to her menstrual periods, when she again assured me that they were unattended by any pain, and that the flow was not at all scanty. I experienced much difficulty in introducing the uterine probe, but after its bulbous extremity had passed the os its subsequent progress to the fundus was perfectly easy. The narrowing was clearly limited to the external opening, the cervical canal and internal os being apparently of normal size.

"The explanation of her condition seemed very plain. The thin lips of the os uteri were sufficiently yielding to permit the menstrual discharge to pass out readily, but from their peculiar shape they closed the opening so completely as to make it impossible, without force for any fluid to enter the uterine canal from the vagina. The case seemed so simple that I was rash enough to promise speedy relief as the result of a slight operation, explaining to her at the same time the nature of the malformation. My suggestions were at once adopted, and shortly afterwards the os was bilaterally slit *secundum artem*. The operation was followed by the introduction of Atlee's dilator every two days until the withdrawal of the slightly-expanded instrument ceased to be attended by any appearance of blood. It was a failure. In less than three months the parts had returned to their original condition, plus a cicatrical induration at the place of the incision.

"A few months afterwards I forcibly dilated the os and lower part of the cervix, — not to the heroic extent sometimes recommended and which I have since practiced in other cases, — but enough to enable me to introduce a No. 24 bougie readily. This also failed. The enlargement effected was only temporary and the parts soon returned to their contracted state.

"Chagrined and disappointed, I had nothing further to propose at the time, and, as I subsequently learned, the patient sought other aid.

"Some months later, when the courageous and hopeful woman again consulted me and asked whether all resources were exhausted in her case, I told her that if she felt willing to submit once more I would make another trial. Accordingly, a few days after the cessation of her next menstrual flow, I removed a circular rim of tissue with scissors, including the lips of the os uteri and extending a third of an inch into the cervical canal, making a beveled wound such as would be made by the mechanic's tool known as a reamer. The operation was done without anæsthesia and caused but little pain. As a precaution against subsequent contraction, I passed a bougie into the cervical canal every third or fourth day until the raw surface was healed. I now believe that this was unnecessary, for there did not appear to be any tendency whatever to undue narrowing, and six weeks after the last introduction of the instrument the os was normally previous.

"At the end of seven months I received a letter from the husband of my patient, who resided out of the city, stating that two menstrual periods had failed to appear, and that they hoped she was pregnant. This hope was realized at last, and the brave woman found her faith and persistence rewarded by the birth of a girl after an easy labor."

LACERATION OF THE CERVIX UTERI.—This lesion is doubtless one of the chief causes of *sterility*. Its importance has hitherto been overlooked. Dr. Emmet was the first to call attention to it, and to him is due nearly everything relating to the surgical treatment, which we now possess. In his classical work * he devotes a large space to its description and treatment, and in this edition I cannot omit mention of it. Dr. Emmet doubts if a woman can give birth to a first child without partial laceration taking place; but slight ones heal rapidly and cause no difficulty afterwards. Lacerations in the median line are most frequent, and those through the anterior lip are more common than those in the posterior one. These heal usually without trouble. But when a laceration in a lateral direction *extends beyond the crown of the cervix*, a condition at once arises which will defeat all the reparative powers of nature. In practice, therefore, we have to deal chiefly with the consequences of lateral lacerations, and the effects are more marked when the lesion is double, than when confined to either side. When the flaps, formed by the laceration, are once separated, they continue to diverge more and more, until the mucous surface of the cervical canal becomes almost completely everted. The angle of the laceration now becomes the seat of erosion, which gradually extends over the everted surface. It is *this* condition which used to be termed "*fissured ulcer, or fissured os uteri.*" These lacerations lead to extensive hypertrophy of the uterus, especially the cervical portion. The surfaces become the seat of profuse cervical leucorrhœa, of a muco-purulent or bloody character. The patient will complain of inability to stand with comfort; has continual backache, pains down her limbs, irritation of the bladder, and general nervous disturbances. Thousands of physicians have treated, and do now treat this condition for *ulceration*, applying caustics, astringents, or the

---
* Principles and Practice of Gynæcology, 1879.

cautery, with none other than temporary relief. Our homœopathic treatment—giving remedies for the symptoms, or making mild and healing applications to the diseased surfaces—will also give temporary relief, without increasing the lesion, as do the severer measures; but a *cure* is impossible, without a proper surgical operation. And if we are treating a patient for *sterility*, the operation must be complete and every way successful, before we can hope to have impregnation occur.

In order that the peculiar appearance of the different forms of this lesion may be readily recognized, I present a cut delineating three varieties of laceration:

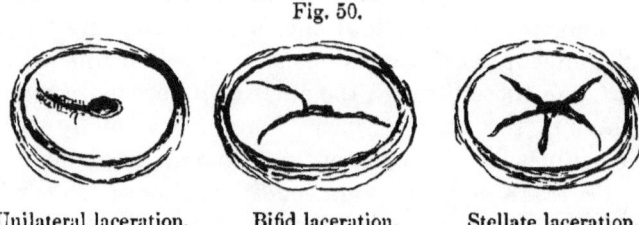

Fig. 50.

Unilateral laceration.　　Bifid laceration.　　Stellate laceration.

In my range of observation I have found the *bilateral* or *double* form of laceration more common than either of the above. This may not be the experience of others. Emmet does not state which variety he considers most common.

Treatment.—Before the operation most cases need a preparatory treatment. The enlarged, congested cervix can be greatly reduced by the use of large, hot water injections, used once or twice a day. These should be used until the tenderness on pressure has disappeared. The water may be slightly medicated with Hamamelis, Calendula, Hydrastis, Iodine, or Arnica. If there is much local irritation (burning and itching), use Borax or Grindelia. In addition, for the double purpose of healing the eroded surfaces and reducing the size of the uterus, also to keep the uterus from sinking into the vagina, *tampons* of absorbent cotton, medicated with Glyceroles* of Iodine, Iodide of Barium, Calendula, Grindelia, Hamamelis, or other

---

* ℨi to the tincture to ℥i of dilute glycerine.

indicated remedies, should be placed against and under the womb every morning and taken out at night, and followed by the hot water enema.

In some cases a *pessary* is absolutely required if the woman be much on her feet. Else the constant dragging and chafing of the uterine cervix will keep up the irritation and prevent the preparatory treatment from doing any good. I prefer for this purpose Jackson's, or Thomas's, bent in a shape suitable to each case, and large enough to *lift the uterus up from the floor of the vagina*, and keep the uterus *anteverted*, for this keeps the flaps together and prevents further separation. The pessary should be so shaped as not to press on any sensitive point. The patient's sense of comfort, and unconsciousness of its presence, is the best guide as to the appropriateness of the instrument.

Emmet recommends the India rubber inflated ring pessary. "If introduced with the flaps of the laceration in contact, and the uterus anteverted, they cannot again separate. Any downward pressure has the tendency to crowd the cervix toward the opening in the ring, while the aperture is not large enough in diameter to allow any portion to become strangulated."—*Emmet.* This pessary should not be large, and should be kept in position by a **T** bandage.

In some cases the flaps are so swollen, that the cervix is strangulated at its junction with the body of the uterus. Added to the stasis of blood, we also find cystic degeneration. (The above cuts show the cysts.) The applications above recommended do not deplete the cervix under such circumstances. It absolutely requires that we should puncture the congested flaps, and the cysts, with *Battle's Spear*. By this means we let out several drams of blood, and empty the cysts, thus removing a great cause of irritation, and the patient will testify to the comfort and relief given by this simple operation.

The *Surgical* treatment, or the *operation* for the cure of the laceration, is fully and minutely described by Emmet, in his great work before alluded to. I can, in this place, only give a brief account of the procedure. It consists mainly in *denuding the internal surface of the flaps* (in double lacerations), *only leaving a broad undenuded tract in the centre from before backward, which* is to form the continuation of the uterine canal from the os. This undenuded portion on each flap is made to correspond with that on the opposite side, and should gradually widen from the edge of the uterine canal towards the outer edge of the divided portion of the cervix. When the two flaps are brought together, the new canal will be trumpet-shaped. As the uterus returns to its normal size, the new canal will become of a natural, uniform diameter throughout.

Emmet prefers the *scissors* to the scalpel, to denude the flaps. He also prefers "*Silver Sutures.*" In the other varieties of laceration the operation is somewhat different, but the general plan the same. For further particulars I must refer to Emmet's work, which should be in the hands of every Gynæcologist.

VII. Vaginal.

Atresia vaginæ.—"This of course forms an obstacle to the ingress of the seminal fluid. It may be *congenital* or *accidental*, more frequently the latter, and oftener the result of a tedious labor, followed by sloughing."—(Sims.)

The vagina may be imperforate at the vulva, or in the course of the canal. The closure may be complete or partial: the former resulting in complete retention of the menses; the latter causing a painful scanty flow, mistaken sometimes for dysmenorrhœa.

The following causes may be enumerated as productive of it:

Arrest of development. Prolonged and difficult labor. Chemical agents locally applied. Mechanical injuries. Sloughing from impaired vitality. Syphilitic or other extensive ulcerations.

Congenital atresia, or arrest of development of the vagina in the fœtus, is a rare anomaly. Aristotle, Celsus, Heister, and Amussat, speak of congenital atresia. Sims says he has seen but one case which might be called congenital. Drs. Helmuth, Franklin, Adams, Danforth, and Comstock inform me they have met with cases. I have met with two cases, occurring in infants: one in a child two years of age, the other a girl of eight years. Both cases I believe to have been caused by arrest of development during fœtal life. The vulvæ were well developed, with a deep sulcus between, but no opening to a vaginal canal could be discovered.

Accidental or acquired occlusion, from any of the above named causes, may occur at the mouth of the vagina, or we may have contraction or closure of the middle portion (of which Meigs and Sims give illustrations); or the upper part of the vagina and the neck of the uterus may be agglutinated together in one dense mass of fibro-cellular tissue, while we may occasionally find a complete obliteration of the canal, from the neck of the bladder quite to the os tincæ.

Treatment.—In *partial* atresia, the menstrual fluids may escape, with much suffering and pain, and the difficulty may never be discovered. Cases are on record in which this condition, occurring in maiden ladies, and even the married, was not discovered for many years, or during life. The abnormal condition may not attract any attention, until an examination for the cause of *sterility* reveals it.

Complete atresia is generally discovered before the woman is married, but not always. The treatment in all cases is to restore the canal, if possible, and keep it open, by the use of a glass dilator, till the newly expanded surfaces are covered with mucous membrane.

Partial atresia is best treated by pressure, or tearing with the fingers, preceded, if necessary, by incisions with scissors. It is best not to use a knife, if it can be avoided, as subsequent contraction and inflammation are more apt to occur than when the tearing process is adopted. Sometimes a cartilaginous band seems to close up the canal. If this cannot be broken by dilatation, the knife must be resorted to. Sims gives some interesting cases.

I have known of two cases, one of congenital occlusion of the vagina, the other a partial occlusion. In the former case, no trace of a vagina could be found, and the operation performed was a delicate one, requiring the most perfect anatomical knowledge. It resulted very successfully: a vaginal canal was made, offering no impediment to coitus, pregnancy resulted, and the woman was safely delivered.

The other case, reported by Dr. W. H. Holcombe, occurred in a married woman aged 35. During an abortion, vaginitis occurred, with almost complete closure of the vagina. When the case was examined three years after the accident, there was a shallow sulcus between the labia, and a very small orifice at the bottom, into which a uterine sound could with difficulty be introduced. A canal was found extending sinuously up to the os uteri. *The woman had become pregnant*, the spermatozoa passing up through this sinus. This woman was safely delivered by Dr. Holcombe, of New Orleans, aided by an eminent surgeon of that city, who operated at the commencement of labor.

CONGENITAL ABSENCE OF VAGINA, according to Scanzoni, rarely, if ever, exists without a simultaneous absence of the uterus and rudimentary development of some of the external organs of generation. If an obliterated vagina be present, it may generally be recognized as a hard fibrous cord, by one finger in the rectum and a sound in the bladder. Should deformity of the external genitals exist, the uterus not be discoverable, and no signs of distress at menstrual epochs show themselves, it may be concluded that the case is one of absence of the vagina, and not of complete atresia. Sims says he has seen five cases of congenital absence of the vagina, and in all of them there was no uterus.

Treatment.—If we diagnose the case as one of absence of the uterus as well as vagina, no treatment can be instituted for the removal of the cause of the sterility. Cases may occur, however, in which the uterus may be present. Amussat reported one to the French Institute, in 1835; and this same bold surgeon has demonstrated that an artificial canal to the uterus might be made.

NON-RETAINING VAGINA.—Dr. Sims was the first surgeon to call attention to this cause of *sterility*. "It has been only about three or four years," he says, "since I found out that some vaginas would not for a moment hold a drop of semen. There are no two vaginas alike. They differ in length, in their various diameters, in their relations with the bladder and rectum, in their course with regard to the pelvian axes, and in their relation with the axis of the uterus. They sometimes refuse to retain the semen when they are very capacious; again when they are too short." Sims relates a singularly interesting case,* where the forcible "reaction of the distended vagina ejected all the semen that did not at once regurgitate in the very act of ejaculation." This he demonstrated to be the case, by inspection of the vagina a few moments after intercourse, when he found that "the vagina did not contain a drop of semen."

In another class of cases, Sims says: "Sometimes the vagina does not retain the semen, even when it is of large proportions. When this is the case, we almost always find the uterus retroverted. I have now but little doubt that, in many cases of retroversion in which I have seen pregnancy follow the rectification of the malposition, the sterile state was due to the fact that the vagina did not retain the semen." Dr. Sims relates several cases illustrating the non-retaining power of the vagina, with retroversion of the uterus, and explains the *modus* of this abnormal action of the vagina.

Treatment.—In the first-mentioned instance, the treatment consists in instructing the parties to so conduct sexual intercourse that a portion of the semen must be left in the vagina; in other words, the semen must not be ejaculated as far as usual into the canal of the vagina. I have several times advised successfully the use of a ball or tampon of cotton to be placed in the vagina immediately upon the conclusion of the act of coition, before the vagina can expel the semen. It is singular that Sims never seems to have thought of this simple device.

In case of *retroversion* with non-retention of semen, the uterus must be placed in normal position, and kept *in situ* by a properly adjusted pessary, which must be worn during coition, as directed in the treatment of retroversion of the uterus. Even in this case, the cotton tampon should be advised, as also in cases where the vagina is too short.

---

* Uterine Surgery, p. 342.

IMPERFORATE HYMEN.—Medical literature contains the records of many cases in which the hymen was so tough as to resist all reasonable efforts at penetration; and very many in which it has been found completely occluded, with retention of the menstrual flow. Sims says he has never met with an example of these conditions. He admits, however, that the hymen may be hermetically sealed, so as to cause retention of the menses and sterility.

Treatment.—All surgeons, at present, caution against the operation, simple as it seems, of opening the hymen by "crucial incision." *i. e.*, if there is more than an ounce or two of retained fluid in the uterus.

The operation should be performed by an exploring needle, leaving the gradual evacuation of the fluid to nature and time. The object of this is to allow the uterus time to contract, as its contents slowly ooze away. If there is not more than an ounce or two of imprisoned fluid, Sims does not think it makes any difference whether we evacuate it suddenly or slowly. If the uterus and vagina hold six or eight ounces, Sims directs to give *Ergot*,* until its specific action is produced on the uterus, and then make a small puncture in the hymen. Dr. Hewitt, one of the latest and best authorities on diseases of women, directs the opening to be made obliquely in the obstructing membrane, giving it a valvular character, so that the fluid will be evacuated *guttatim*. An abdominal bandage should be worn till the fluid is all evacuated.

In cases where the hymen is so nearly closed as to admit the escape of the menses drop by drop, and the orifice so small as to prevent the entrance of the semen, it had better be ruptured by pressure, if possible, than by any cutting instrument, as in occlusion of the vagina.

VAGINISMUS.—This singular affection is much oftener a cause of *Sterility* than has been supposed. It was first pointed out by Dr. Burns, of Glasgow, and more lately, by Simpson, Scanzoni, Sims, and others; also by our own Helmuth,† who first called

---

* Caulophyllin, perhaps, and would be as good. (H.)
† North American Journal, 1864, p. 64.

attention to it in our journal, and reported a case treated successfully.

Definition.—" By the term vaginismus I mean an excessive hyperæsthesia of the hymen and vulvar outlet, associated with such involuntary spasmodic contraction of the sphincter vaginæ as to prevent coition. This irritable spasmodic action is produced by the slightest touch; often the touch of a camelshair pencil, or fine feather, will produce such agony as to cause the patient to shriek out, complaining at the same time that the pain is that of thrusting a sharp knife into the sensitive part. This is worse in some than in others. In a very large majority the pain and spasm conjoined are so great as to preclude the possibility of sexual intercourse. In some instances it will be borne occasionally, notwithstanding the intolerable suffering; while in others it will be wholly abandoned, even after the act has been repeatedly, and, as it were, perfectly performed" (Sims).

The recognized causes are: The hysterical diathesis; excoriations or fissures at the vulva; irritable tubercle of the meatus; chronic metritis or vaginitis; pustular or vesicular eruptions on the vulva; neuromata.

Diagnosis.—There is no other affection with which it can be confounded. All that will be necessary to decide concerning it will be whether it is an idiopathic or symptomatic disorder.

Course and Duration.—In its course it may be unlimited. Cases are recorded in which it lasted for twenty-five or thirty years. In some rare cases it may be removed by medicines acting on the nervous system, or pass away without treatment.

Prognosis.—" From personal experience," remarks Dr. Sims, " I can confidently assert that I know of no disease capable of producing so much unhappiness to both parties to the marriage contract; and I am happy to state that I know of no serious trouble that can be so easily, so safely, and so certainly cured."

Treatment.—The surgical treatment is given for this affection by Dr. Franklin and Dr. Helmuth, in their works on Surgery.

Guernsey ignores the disease altogether, or includes it in the chapter on "spasms, cramps, constrictions, and neuralgia of the vagina," although he does not therein describe the true vaginismus of which we are treating. With characteristic boldness he asserts that "these strictures of the vagina may be cured without artificial dilatation, by the aid of the homœopathically indicated remedy." This assertion, however, is not supported by a single clinical fact.

The rational treatment of vaginismus should be conducted as follows: First ascertain, if possible the *cause* of the affection, and whether it is idiopathic or symptomatic.

If it arises from the *hysterical diathesis*, those remedies indicated in hysteria, and possessing the general and local symptoms, should be selected. Among these the most useful are: *Platina, Belladonna* (or *Atropin*), *Conium, Asafatida, Hamamelis,* and *Gelseminum*.

The introduction of suppositories of cocoa butter, medicated with *Conium, Belladonna, Chloral, Atropin, Iodoform,** or *Gelseminum,* as they are indicated, will aid the internal administration of the remedy. At the same time a glass tube, known as "Sims's vaginal dilator," should be gently inserted into the vagina and kept there for as many hours a day as practicable. Its presence tends to benumb the nervous sensibility, and produce a tolerance of foreign bodies. During the treatment the patient should live apart from her husband.

If excoriations, ulcers, or fissures at the vulva are present, Nitric acid, Ignatia, Graphites, and *Sulphur,* may be given internally, but the best results will be gained by applying the fuming nitric acid to the fissures or ulcers.

One of the former remedies may be alternated with one of the latter with good effect; at the same time the *Graphites cerate, Calendula, Glycerin,* or *Glycerole of aloes,* may be applied locally. The *Aloes glycerole* is superior to any other known remedy in the treatment of irritable and obstinate rhagades, or fissures, occurring anywhere. An ointment of Chloral hydrate or Iodoform, will also give relief (gr. x to Vaseline ʒj).

If an *irritable tubercle of the meatus or vagina* exists, a surgical

---

* A suppository, made of equal parts of *Tannin* and *Iodoform,* is almost destitute of the abominable odor of the latter, but may not act as well in this disease.

operation is the only resort. It should be hooked up with a tenaculum and cut out, and immediately the peculiar sensitiveness of the part will disappear.

Dr. Baker Brown reports many cases of vaginismus cured by excising the enlarged and irritable clitoris. (See observations on that disease below.)

Pustular eruptions of the vulva require the use of *Tartar emetic, Sabina, Croton tig., Arsenic.*

Vesicular eruptions call for *Rhus tox., rad.* or *ven.,* and *Euphorbia. Urticaria of the vulva* requires *Apis mel., Urtica urens. Graphites, Arsenic. Neuromata* require to be removed by the knife or curved scissors.

I have recently read the report of a case of *vaginismus* cured by the same process which has been adopted successfully in the treatment of fissure or irritable ulcer of the rectum, namely, *forcible dilatation.* The reporter stated that he introduced his thumbs into the vagina, and forcibly stretched it as much as possible, this operation was repeated once or twice a week, and resulted in a cure in six weeks. As his patient was not under the influence of an anæsthetic, the case could not have been a severe one. I would suggest that the use of Barnes's dilators, or the inflatable pessary, or my expanding speculum would give better results.

Since the first edition of this work was issued, I have invented an expanding and self-retaining speculum. This instrument was devised to meet a need often felt by physicians doing a large gynæcological practice. All the ordinary specula always cause great pain and distress in such conditions as vaginismus, contraction of the vagina, or in young unmarried women. The very act of introducing the ordinary specula causes suffering, and when introduced the distress is such that the patient cannot retain them long enough to allow the operator to explore the uterus, or make the necessary applications. If a cylindrical speculum, small enough to be readily introduced, is used, the expulsive efforts of the vagina make it difficult to keep it in place. Besides, the orifice of such a specula is so small that no satisfactory examination can be made.

Feeling the great need of a small expanding, self-retaining

speculum, in my own practice, and not finding such a one in the market, I devised the instrument represented in the following outline cut.

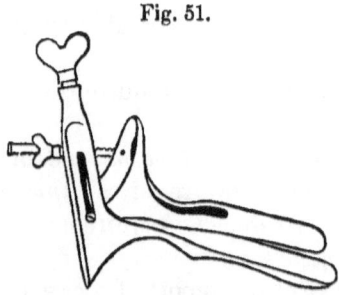

Fig. 51.

While bearing a general resemblance to my bivalve expanding speculum, which has met with such excellent reception, it presents some evident improvements:

1. It is about *half* the size, in weight; *two-thirds* the width in the blades and fixed opening, and *three-quarters* as long as my large speculum.

2. There is a *depression at the base of each blade* which, when introduced, allows the sphincter, at the orifice of the vagina, to sink into them, and prevent the extension of the instrument. This is an important improvement and does away with that distressing feeling of stretching and tearing which all other specula causes.

With the exercise of some tact cases of vaginismus or irritable vagina can be treated with this instrument without other surgical procedure. When the instrument is introduced, slowly and carefully elevate the upper blade a little by placing the finger behind its upward projection, and drawing it gently forward, or turning the nut on the screw which projects from the upper blade. (It is necessary in some cases in order to bring the os in view, to depress the lower blade by introducing the fore-finger of the left hand, before we elevate the upper blade). Then carefully turn the screw which separates the two blades at the outer extremity. By means of this alternating action the muscles of the ostium vagina are soon paralyzed, and the instrument can be retained a considerable time without much suffering, allowing the cervix or internal uterus to be sufficiently explored.

I think this instrument can also be used advantageously as a *rectal speculum*, for lately, when treating an obstinate case of uterine disorder in a very young woman, which I believed to be due to some disorder of the rectum, on attempting to make an examination I found a condition corresponding to vaginismus—due to a small *fissure*. It occurred to me to use the small speculum, which I accomplished after introducing the finger (the patient being under the influence of Ether). When introduced I slowly separated the two blades, paralyzing the sphincter, and tearing the base of the fissure. This resulted in a rapid cure of both the rectal and uterine trouble.

Before describing the *surgical treatment* of vaginismus, it ought to be mentioned that several cases are on record where coition was performed while the wife was under the influence of *chloroform* or *ether*, and that conception resulted from such intercourse. In some of these cases the vaginismus disappeared after confinement; in others no improvement resulted.

Dr. T. G. Thomas of New York, related to Dr. Sims a case in which pregnancy followed coition under the above circumstances.

Dr. Sims relates a case in which two conceptions occurred in a state of complete anæsthesia, followed by a labor at full term, and a miscarriage, but without curing the vaginismus.

Treatment.—The Treatment consists in the removal of the hymen, the incision of the vaginal orifice, and subsequent dilatation. The last is useless without the first two, but is essential to easy and perfect success with. " I usually make two operations, but it may be done at once."*

" The patient being under the influence of *ether*, and placed on the back upon a table, the remains of the hymen are entirely excised by a pair of curved scissors. The slight hæmorrhage resulting from this will soon cease under the application of a compress wet with ice-water, or a solution of *persulphate of iron*.

---

* Sims's Uterine Surgery, page 326.

"The index and middle fingers of the left hand are then passed into the vagina, so as to put the fourchette on the stretch. By means of a scalpel, a deep incision is then made on the right of the median line, terminating at the raphe of the perinæum. A similar incision is then made on the other side, the two being united at the raphe, and extended to the perineal integument, and through its upper border. Each of these incisions will extend from about half an inch above the upper border of the sphincter to the perineal raphe, thus passing across the muscle, and measuring nearly two inches. They should pass over the sphincter muscle, but not *entirely* through it.

"After this, the vaginal dilator is placed in the canal, either immediately or in about twenty-four hours, and worn for two hours in the morning, and three or four in the evening, according to the tolerance for it which is manifested. The vaginal dilator is three inches long, slightly conical, open at one end, and closed at the other, and varying in size from an inch to an inch and a half in diameter. This instrument is kept in place by a **T** bandage, and should be worn for two or three weeks."

LEUCORRHŒA (VAGINITIS).—Vaginal leucorrhœa is always the result of acute or chronic vaginitis. It may be specific, arising from gonorrhœa, catarrhal, or caused by various kinds of irritation.

"I do not know," says Sims, "that vaginitis, properly speaking, is absolutely opposed to the vitality of the spermatozoa. It appears to be the *character*, and not the quantity, of the vaginal secretion, that kills the spermatozoa."

"The secretion from the vagina should be slightly acid; if it is very acid it kills the spermatozoa instantly. I have seen many cases in which they were all dead within five or six minutes after coition. In all these cases, the vaginal mucus was by no means abundant, but the surface of the vagina always had a reddish look, and its papillæ were prominent. By simply inspecting the surface of the vagina, and testing the degree of acidity with litmus-paper, I have sometimes been able to say that the vaginal mucus would poison the spermatozoa. The blue litmus should be slowly turned to a faint pink, when the secretion is normal; but when it is abnormal, the litmus-paper turns quickly to a deeper pink color."

Treatment.—The treatment of leucorrhœa or vaginitis, in cases of *sterility*, should be *chemical* or *medicinal*, or both.

The *chemical* consists in so changing, temporarily, the abnormally *acid* condition of the secretion, that it will not be fatal to the life of the spermatozoa.

"I have seen," says Dr. Sims, "conception twice (occur) when the vaginal mucus poisoned the spermatozoa. One was remedied by slightly alkaline washes, used before sexual congress. In the other it occurred in this way: A lady, aged twenty-eight, was married six years without issue. She had a contracted os. It was incised, but she did not conceive. She had indurated cervix, the consequence of cystic disease. For this she was under treatment nearly two months. It was cured, and her husband came to take her home. Wishing to see the character of the semen, I examined the vaginal mucus four or five hours after coition. The spermatozoa were all dead. On the next day, I examined them in five or six minutes afterwards, and could not find one alive. I then placed in the vagina a small tampon, moistened with a little *glycerin*, which held in solution some of the *bicarbonate of soda* (20 grs. to ʒj). This application was repeated on the next day. The cotton was tied with a string, for its removal. This was worn from about two o'clock P.M., till eight the next morning. Its removal was followed by connection. Living spermatozoa were afterwards found in the greatest abundance; indeed, there were no dead ones at all. Conception dated from that moment, being just two days before the expected return of the menses, which, however, did not occur. There had been no sexual intercourse for nearly two months before."

According to Kolliker, the *phosphate of soda* is peculiarly favorable to the movements of spermatozoa; and this would probably be superior to the *carbonate* as an application in extremely acid conditions of the vaginal mucus.

The *medicinal* treatment of *acute* vaginitis is best conducted by the internal administration of *Aconite, Cantharis, Copaiva, Cannabis sat., Clematis, Pulsatilla, Mercurius, Thuja,* and *Sabina.* For *chronic* vaginitis: *Merc. cor., Sepia, Calcarea, Arsenicum, Senecio, Kreosotum,* and *Sulphur.* Besides the remedies mentioned, I have found from recent experience, that in acute vaginitis, with local *heat,* etc., enemas of *Gelsemium, Grindelia* or *Hamamelis,* together with their internal use, are capable of removing the inflammation sooner than any other treatment.

In chronic vaginitis. my experience is strongly in favor of the use of tampons of cotton, saturated with dilute Glyceroles (ʒj to ʒj) of *Calendula, Muriate of Hydrastia, Myrica, Grindelia Bismuth, Chloral,* or *Borax.*

Diseases of the Clitoris.

That some congenital and acquired diseases of the clitoris indirectly cause *sterility* is placed beyond a doubt by the investigations of Baker Brown, whose success in curing not only sterility but insanity, epilepsy, catalepsy, and hysteria, should have prevented the senseless folly of the members of the "London Obstetrical Society," who expelled him from that Society for reasons which will be utterly inexplicable to future surgeons and physicians.

In Dr. Brown's *Surgical Diseases of Women* he devotes a chapter to "Irritation and Hypertrophy of the Clitoris," and writes:

"Enlargement of the clitoris, sometimes accompanied by a degree of induration approaching that of cartilage, at others by a relaxed, flabby state of the tissues, and always attended by abnormal irritability, is a condition of more frequent occurrence, I believe, than the majority of medical men suspect, and is for the most part brought on by self-abuse. The radical cure of the habit is, however, fortunately in our hands. Long-continued irritation of the clitoris figures among the causes of *sterility,* for, besides its constitutional effects, the habit acts locally on the functions of the womb, either in the same way, we may presume as does excessive venery, or by inducing displacement of the organ."

The following symptoms are noted by Brown as being pathognomonic of this habit and condition: "Melancholy; restlessness; excited and retiring; listlessness, and indifferent to social influences and domestic life; strange fanciful appetite, dyspepsia; pain in head and down the spine; pains more or less constant in the small of the back, or on either side in the lumbar region; wasting of the face and muscles generally; skin dry and hard, or cold and clammy; pupils dilated; hands clammy; often a hard cordlike pulse; irregularity of the uterine function; amenorrhœa, or too frequent menses; dysmenorrhœa and leucorrhœa. If married, the woman has distaste for marital intercourse, and very frequently either *sterility*, or a tendency to abort in the early months of pregnancy.

"On examination of the external genital organs there will be found a straight and hirsute growth, a depression in the centre of the perinæum, a peculiar follicular secretion, an alteration in the structure of the parts, mucous membrane taking on the character of skin, and muscle having become hypertrophied and generally tending toward a fibrous or cartilaginous degeneration."

Surgical Treatment.—" When I have decided that my patient is a fit subject for surgical treatment, I at once proceed to operate, after the ordinary preliminary measures of a warm bath and clearance of the portal circulation. The patient having been placed *completely* under the influence of chloroform, the clitoris is freely excised either by scissors or knife. I always prefer the scissors. The wound is then firmly plugged with a graduated compress of lint, and a pad well secured by a **T** bandage. A grain of Opium is introduced into the rectum, the patient placed in bed, and most carefully watched by a nurse to prevent hæmorrhage by any disturbance of the dressing. The diet must be unstimulating, and consist of milk, farinaceous food, fish, and occasionally chicken. The strictest quiet must be enjoined, and the attention of relatives, if possible, avoided, so that the moral influence of the medical attendant and nurse may be uninterruptedly maintained." (Brown.)

Dr. Brown says: "A month is generally required for the healing of the wound, at the end of which time it is difficult for the uninformed or non-medical to discern any trace of any operation. The rapid improvement after the operation is most marked. The woman soon recovers excellent health, mentally and physically, and a healthy pregnancy results. He asserts that this operation no more 'unsexes' the woman than an operation for piles. On the contrary, healthy and natural sexual appetite and pleasure take the place of the abnormal appetite."

Medicinal Treatment.—It is a very hazardous and delicate matter for a physician to propose this operation, be it ever so necessary, owing to the popular and even professional prejudice. We are, therefore, driven to our local and internal medicinal remedies, by which we may hope to accomplish results approximating excision.

Abnormal irritability of the clitoris may, however, be due not to any pernicious habit, but is often a congenital disorder of the pubic nerve which supplies that organ. We may not be enabled by medicinal influences to permanently remove abnormal irritability of the clitoris when it has been acquired, or when that organ has become enlarged or otherwise changed in structure; but we may be able to palliate the irritation to such an extent that *conception* can take place, and we may also be able, by the use of appropriate remedies, to carry the pregnancy to a final termination.

In cases of congenital irritability, remedies have a much more permanent effect, especially if masturbation has not been contracted, and the clitoris remain normal in structure. In such cases pregnancy usually arrests the irritability, and after childbirth it disappears altogether.

There are but few medicines upon which we can rely in this disorder, and in order to be successful with them we must prescribe them in material doses until their primary physiological effects are established.

These medicines are (1) the Bromines, especially the Monobromate of camphor, (2) Picric acid and Hydrobromic acid, (3) Conium, (4) Ferrocyanuret of potassium and Bromide of iron.

Monobromated camphor has a prompt and decided action as a sedative to the pubic nerve and the vaso-motor nervous system generally. It not only calms the general nervous system, but those of the genital organs especially. I have used it with such success for several years that I can confidently recommend it.

ILLUSTRATIVE CASE.

Sterility, Congenital Abnormal Irritability of the Clitoris Cured.—Was applied to by Mrs. —— for the removal of sterility. Had been married six years, without conception. She was a woman of large physique, bilious temperament; had internal hæmorrhoids, and chronic constipation, dysmenorrhœa, and leucorrhœa. *Examination* showed some follicular inflammation of os uteri, chronic vaginitis, and some vaginismus. The vulva was rough and the parts presented the appearance described by Dr. Baker Brown, above quoted, except that the clitoris was *not* enlarged. It was, however, highly irritable as accidental touching showed. Although I suspected something abnormal, from the spasmodic contractions of the vagina, and the excessive transparent mucous secretion poured out by the cervical glands and those of the ostium vaginæ, I did not, from motives of delicacy, ask any question on that point. Treatment for a few weeks removed the follicular ulceration, the vaginitis, and the constipation. In a conversation with her husband, he asked me if a state of almost constant sexual excitement would not induce sterility. I answered in the affirmative, when he requested me to consult with her on that subject. On personal inquiry she related that she had been the subject of abnormal irritability of the external genitals since her earliest childhood, "as long ago as she could remember anything." To get relief she was obliged to resort to rubbing and friction, and this would bring on orgasm. This led to the habit of masturbation, which continued till she was married, when she hoped it would cease, but was disappointed, for her husband could not satisfy the unnatural irritation. She had vainly striven to control it, and although she was evidently a woman of strong mind and good principles she failed because the *mind* was not at fault. When she resisted successfully during waking hours, she involuntarily produced orgasm by rubbing *during sleep*. She suffered much from occipital headache, extending down the spine; had wild and unpleasant dreams, and a "wild feeling in her brain that alarmed her." Gelsemium 1^x was prescribed with but little effect, as also was Bromide of potassa, which only gave her better sleep. I then gave her capsules of Camphor monobromide each containing $2\frac{1}{2}$ grains, of which she took three a day. In a few days she reported herself much better in every respect, and in three weeks the pain in the head had ceased, and sexual functions were about normal. The next month she missed her menstrual period, the first time in her life, and soon found she was pregnant. During pregnancy she had to resort occasionally to the medicine, but went to full term and gave birth to a healthy child. She informs me that since the birth of the child, now over a year, there has been no return of the former unnatural excitement.

I have since used the Camphor monobromide in another similar case with like result. With the Bromide of ammomium I once cured a case presenting the above features, with this exception,—the abnormal irritation occurred only before, during, and after the menses; the dose was ten grains, three times a day.

Picric acid differs from the above in being primarily homœopathic to the above-mentioned normal condition, when given in small doses; *i. e.*, small doses *cause* sexual erethism; large doses produce a morbid sexual depression. Dr. L. B. Couch* says he has cured several cases of self-abuse by this medicine. In one case reported, the cure was made by the 30th. This proves my theory of dose to be true. In obstinate cases Dr. Couch says he should not hesitate to give large doses (10 grains of the 1^x trit.).

Hydrobromic acid is said to possess nearly all the properties of the bromides, and as it is very pleasant to the taste made into a "lemonade," and never disagrees with the stomach, it may prove a useful remedy. Drop doses, if pure, should represent grain doses of the bromides of soda, lime, and potassa. In cerebral congestion I get good effects from 15 to 25-drop doses.

Conium will be treated of in a chapter further on. I will

---

* Homœopathic Times, April, 1878.

only say here that in abnormal irritation of the clitoris, I have used Squibb's fluid extract of Conium, in doses of 5 to 15 drops once or twice a day with good effects, in a few cases.

Ferrocyanuret of potassium and *Bromide of iron* will prove excellent remedies in this affection, in chlorotic or anæmic women subject to cardiac and intercostal neuralgia, when the menses are pale and too frequent. They enrich the blood, while they lower in a marked degree the sexual power and appetite. This is especially the case with the Cyanuret.*

---

* See vol. ii, New Remedies (Therapeutics).

CHAPTER VIII.

Renal.

Diabetes.—"During the course of this disease," says Dr. Morgan,* "the power and function of reproduction are more or less suspended in both sexes, but returns as vigorously as ever on the removal of the diabetes.

Diabetes mellitus, and perhaps diabetes insipidus may be a cause of sterility, and not be recognized as such by the physician, who may be treating some uterine trouble which he supposes is the real cause.

Treatment.—The most important remedies for this disease happen to be at the same time some of our most valuable remedies for sterility, viz.: *Acidum phosphoricum, Mercurius, Helonias, Arsenic, Baryta carb., Conium, Digitalis, Eupatorium purp., Kali carb., Uranium nitrate, Terebinth., Acidum mur.* (For the special indications for these and many other medicines, see Dr. Morgan's work.)

I have cured two cases of diabetes in women with Helonias, in both of which impotence and sterility were concomitants. After their recovery sexual power and desire returned, and they soon became pregnant. The 1^x dilution was used, ten to fifteen drops four times a day. Those who wish to consult the important clinical and pathogenetic history of this remedy will find it in my *Therapeutics of New Remedies*, 4th edition.

Lycopus virginicus has been found curative in several cases of diabetes.†

Secale has been used very successfully in *polyuria* (D. insipidus) after the failure of many other medicines. It was given in doses of fifteen to thirty drops of Squibb's fl. ext., three times a day. It also cured the attending impotence.

---

* Diabetes, etc., by Dr. William Morgan, Hom. Pub. Co., London, 1878.
† See August number North Amer. Journal of Homœopathy, 1878.

CHAPTER IX.

VESICAL.

IRRITABLE BLADDER AND URETHRA.—These two distressing and obstinate disorders may become very prominent causes of sterility, by the reflex irritation transmitted to the uterus and vagina.

I have found that vesical catarrh is nearly always attended by vaginal and uterine catarrh, and the cystic disease may have been the first to appear. The nervous irritation is also transmitted to the uterus, causing spasmodic or neuralgic dysmenorrhœa.

There are several varieties of cystic and urethral disease, and a brief mention only will be made.

Chronic cystitis, or *vesical catarrh*, is one of the most painful and annoying affections with which a woman can be afflicted.

It commences usually with a simple cystitis from a cold, especially from exposure during the menses. At first the discharge is simple mucus mixed with the urine, but as the disease advances it may become bloody, and is attended with severe dysuria, and lancinating, burning, and spasmodic pains like tenesmus.

In the chronic stage the dysuria is intense, the urging to urinate very frequent, attended by vesical tenesmus, the sleep is broken by frequent calls to urinate, the urine is scanty and high-colored, containing a large proportion of pus, muco-pus, blood, epithelium, and in the worst cases, shreds of mucous membrane from erosion and sloughing. The region of the bladder is sensitive to the touch; walking or riding is painful, and finally hectic fever sets in. In some cases the urine is intensely fetid, owing to rapid decomposition, or excessive quantity of phosphates, especially if the posterior wall of the bladder becomes prolapsed, and pouches into the rectum. Not only does this affection cause sterility by reflex irritation of the uterus, but it renders fruitful marital congress almost impossible, on account of the pain and irritation it causes, and because of the necessity of getting up to urinate immediately after the act, when the tenesmus of the bladder is attended by sympathetic tenesmus of the uterus and vagina, expelling the seminal fluid from those organs.

Not only is it necessary to cure the disease before conception can take place, but even should conception occur, the woman would be in a very pitiful condition, for pregnancy, if not cut short by a miscarriage (a common result), would be a continued scene of indescribable suffering.

Treatment.—The diet of the patient must first be regulated. All stimulating beverages containing alcohol should be prohibited. All foods containing spices, pepper, mustard, horseradish, etc., should be avoided. Milk is the best article of food, with bread and meat. Walking and riding should be indulged in with great moderation. Coition should be very rare, if at all, until a cure is effected. The feet and lower extremities should be dressed warmly. Damp feet should be avoided by wearing wollen hose, thick shoes with cork soles, or voltaic insoles.

The medicines most efficacious in *acute* cases are Cannabis, Equisetum, Cantharis, Pulsatilla, Capsicum, Galium, Spirits nitri dulc., etc.

In *chronic* cases the most successful remedies are *Galium, Chimaphila, Mitchella, Uva ursi, Pareira brava, Barosma, Grindelia, Monobromate of camphor, Santonin, Benzoic acid* and the *Benzoate of lithia, Thuja, Terebinth., Copaiva,* and *Cubebs*. With these medicines, if carefully selected, nearly all cases of vesical catarrh can be cured. Some of them, namely, Galium, Chimaphila, Mitchella, and Uva ursi, act best when given in infusion, cool or warm, prepared in the proportion of ʒj to ʒss of the herb to a pint of water, a wineglassful taken every few hours. Barosma and Pareira act best in Syrup bal. Peru (ʒj to ʒiv of the Syrup, a teaspoonful every few hours). Thuja and Terebinth. in the 2^x dilution. The others in the 1^x trituration, a few grains every two or four hours.

The *topical* treatment of this disease is indispensable in

many cases. The medicines in filtered solutions should be thrown into the bladder, by means of a syringe directly, or through a short or long catheter. As a preliminary to this treatment the urethra will often have to be dilated. (See Diseases of Urethra.) The best medicines for topical use are Hydrastis, Grindelia, Copaiva, Terebinth., Thuja, Salicylate of soda, Bicarbonate of soda, or Acidum muriaticum.

Hydrastis, when the sediment is tough, ropy, and mucopurulent (ʒj of Merrill's fluid hydrastis, or gr. v of the muriate or sulphate of hydrastia to a pint of warm water).

Copaiva, Terebinth., Grindelia, and Thuja, in the form of medicated waters, made as follows: ʒj or ij is rubbed up with carbonate of magnesia, then added to a quart of water and filtered. Inject half a pint or more daily.

It is best to thoroughly wash out the bladder before the medicated enema, by injecting simply warm water, or very dilute solutions of muriatic acid if the urine is alkaline; if the urine is acid, sodæ bicarb.* Some cases have been cured by injecting *healthy* urine taken directly from another person, and thrown into the diseased bladder. Salicylate of soda, or Salicylic acid, used as enemas, have cured many cases when the urine was intensely fetid and contained vibriones.† (In cases of prolapsus of the bladder I have applied a Thomas's anteversion pessary with great success.) The most intractable case I ever saw was cured by the latter. The woman had been a victim of this disease for years. She was married and childless. The only relief she obtained was by injecting warm water several times a day. She had adopted the plan of her own notion. I found the urethra already dilated, so that I could introduce my index finger. I prescribed the following:

R.—Salicylic acid, ... ʒj.
Glycerin, ... ℥j.
Water, ... 1 quart.

---

* The addition of a grain of Morphia, or $\frac{1}{20}$ grain of Atropin, is an excellent palliative of pain in very bad cases.

† R.—Salicylic acid, ... gr. xx.
Borax, ... gr. xxx.
Water, ... 1 quart.
                Or,
R.—Salicylate of soda, ... ʒss.
Water, ... 1 quart.

Four ounces of this was thrown into the bladder, previously emptied and washed out with warm water, and repeated several times daily.

The next most successful cure was made by giving internally, in teaspoonful doses every four hours, a mixture of equal parts of Tincture of Chimaphila and Syrup Balsam of Peru. She was cured in two weeks, after many of the above medicines, even Chimaphila in drop doses of the tincture, had been used unsuccessfully. Both these women became pregnant soon after the cure. (Since the above was written I have cured a severe case with Grindelia robusta, given internally, and applied topically, prepared as a "medicated water.")

Surgical Treatment.—Many gynæcologists and surgeons are now treating chronic cystitis, irritable bladder, and even vesical catarrh, by means of dilatation of the urethra. It is claimed with good reason that dilation paralyzes the irritable sphincter and ruptures the circular fibres, and allows the diseased bladder to get well. Dr. Skene has just issued a valuable work on "*Diseases of the Bladder in Women,*" which contains all our present information on this subject.

CHAPTER X.

Urethral.

IRRITABLE URETHRA, or CHRONIC CATARRHAL URETHRITIS may become a cause of sterility, equally with the similar affection of the bladder. In this disease the pain and suffering during and after urination and coition, are more intense than in chronic cystitis. It so wears upon the system that women become nervous, irritable, and almost insane from the constant irritation, the loss of sleep, and the peculiar sufferings. In cases of true irritability of the urethra, no discharge is observed, and a close examination fails to detect any abnormal growths in the canal or at its mouth. It seems to be as purely a nervous affection as are some severe cases of vaginismus. In catarrhal urethritis, which may arise from repeated acute attacks, or from gonorrhœa, structural changes often occur, namely, thickening of the walls, fungous growths, ulceration, occlusion, or paralysis.

Treatment.—The medicinal treatment is far less satisfactory than for chronic cystitis, but the same remedies are indicated. The topical treatment is the most important. In the neuralgic variety, *atropia** applied to the whole length of the canal by means of a probe wrapped with cotton, or *carbolic acid*† may be useful in some cases. But the most successful treatment is by means of *dilatation*. This operation is equally efficacious in nearly all diseases of the female urethra, and is adopted by all our best surgeons.

This simple operation is so important and useful that I will give the most approved methods now adopted. Sir Astley Cooper first proposed and practiced it for the purpose of removing stone from the bladder, but he used the *slow* method, by means of sponge, or sea-tangle tents. But this method has

---

* Atropia, gr. i; Glycerin, ℥j.
† Carbolic acid, gr. v; Glycerin, ℥j.

recently been supplanted by the equally safe and thorough and much less tedious method of *rapid and forcible dilatation*. As regards the instruments used, Simon employs graduated bougies or special bivalvular specula, first incising the border of the meatus. Noeggerath dilates with graduated steel bougies and the little and index fingers. Thomas with a forceps similar to a dressing forceps. Munde says Goodwillie's wire-spring nasal speculum is a good instrument. Dr. Ball's instruments, which he uses for forcible and rapid dilatation of the cervix uteri, might be useful. Any of the forcible dilators of the cervical canal will answer; in many cases the finger alone will do the work, for the urethra is often dilated somewhat by the disease. In all cases the dilatation occupies but a few minutes or seconds, and is *never* followed by permanent incontinence of urine; all operators agree in this statement. Heath says the urethra *splits* from beneath the pubis, and that the incontinence rarely lasts longer than twenty-four hours. The operation is always performed under anæsthesia. The benefit is instantaneous. The *American Journal of Obstetrics*, and other journals during the last year or two, have contained many reports wherein this operation has permanently cured old and hitherto intractable cases, and restored the patients to health.

CARUNCLES OF THE URETHRA.—This is one of the most common of urethral disorders in the female sex. Dr. Goodell* describes them as intensely vascular and exquisitely sensitive growths, occurring more frequently in married women; they do not appear to be due to gonorrhœa, uncleanly habits, or frequent coition, but are usually accompanied, perhaps preceded, by uterine diseases. Although at times perfectly painless, they are often the cause of the most excruciating pelvic and vesical pains, and in the course of time bring about alarming symptoms of general debility, nervousness, and depression, apparently quite out of proportion to the size of the tumor. The caruncles may occupy any portion of the urethra, and are the most difficult of treatment when situated near the bladder and sessile in their attachment.

---

* Affections of the Female Urethra, Philadelphia Medical Times, 1874.

Treatment.—It is possible that medicines may be useful in some cases, but doubtful if they will more than palliate. *Thuja* is the best indicated. Goodell advises their destruction by means of chromic acid, nitric acid, or carbolic acid, applied full strength, or they may be removed by the scissors (followed by the immediate application of fuming nitric acid, or the actual cautery, to check the usual profuse hæmorrhage), or by the galvano-caustic loop, or its less expensive substitute, a red-hot knitting-needle or hair-pin. Forcible dilatation of the urethra will sometimes be found necessary to prevent their return and effect a permanent cure.

GRANULAR EROSION OF THE MUCOUS MEMBRANE is a very painful affection, but generally relieved by the application of undiluted carbolic acid on a fine stick, once a week, followed by the introduction of Olive oil or Cosmoline.

FISSURE OF THE URETHRA is a rare and obscure form of urethral disease. It occurs near the neck of the bladder. It is readily cured by hyperdilatation and touching with lunar caustic or nitric acid.

VESICO-VAGINAL FISTULA is a cause of sterility. For details of the operation for its removal consult the standard works on surgery.

VASCULAR TUMOR OF THE MEATUS.—" Few diseases of trifling magnitude occasion more distress than a vascular excrescence, varying in size from a large pin's head to a horse bean, which is sometimes found growing from the female urethra. Its exquisite sensibility shows it to be as well supplied with nerves as with bloodvessels. The tumor sometimes arises from the projection which generally exists around the orifice of the meatus, but it frequently grows from the internal surface. *The tenderness of the part is so great as not to allow of sexual intercourse, and it may thus become indirectly a cause of sterility.*" (Dr. Baker Brown; *Surgical Diseases of Women.*)

Diagnosis.—An exquisite sensibility of the part is a leading symptom of the disease. The patient complains of excessive pain in micturition, in coition, and from the slightest pressure on it. Upon separating the vulva there will appear a small tumor, of a florid scarlet color, resembling arterial blood. It springs from the orifice or just within the orifice of the urethra. It easily bleeds on rough handling. Its surface is somewhat granulated. There are sometimes several of these tumors.

Treatment—" Ligatures," says Baker Brown, " are useless as they cannot be made to include the whole of the diseased mass, nor can they be tied with any degree of force, so as to strangle it without exciting inflammation, as, in order to effect a cure, some proportion of the mucous membrane should be included. The best practice is to excise the tumor with a pair of scissors, taking care to remove not only the excrescence itself, but also that small portion of mucous membrane from which it grows, a fine pair of forceps being used to take hold of it. To the wound thus made Nitric acid should be applied on a piece of stick pointed like a pencil, the parts around being filled with lint previously soaked with a strong solution of Carbonate of soda." Dr. Brigham, however, says that they are quickest cured by touching their extremity with the actual cautery. I have cured small ones by means of Persulphate of iron, and Nitrate of sanguinaria.

CHAPTER XI.

Rectal.

"Diseases of the rectum," says Dr. Baker Brown, "will produce sterility. The rectum and uterus are both supplied by vessels and nerves from the same source; and, therefore, disease in one organ must interfere with the other. When a female is suffering from *bleeding hæmorrhoids* during the menstrual period, a diminished supply of blood is sent to the uterus, and its mucous membrane will not undergo those normal changes necessary for the reception of the impregnated ovum. The same observations apply to *prolapsus ani*, with loss of blood at every defecation."

In Volumes I and II of the *American Journal of Obstetrics* appears a series of papers relating to the rectal causes of sterility and uterine diseases, by Dr. H. R. Storer, of Boston. They will amply repay perusal by the physician who desires to become thoroughly acquainted with this subject.

Prof. T. G. Comstock, of St. Louis, communicates the following case:

'Fissure of the Anus causing Sterility.—"Mrs. W., aged 31, married ten years, was never pregnant. Had suffered for seven years with what she supposed was piles. She had been treated by many physicians, who seem to have been mostly in accord as to the diagnosis, they supposing it to be internal piles. She had never allowed any examination by the speculum, until she came under my charge. I gave her case a careful investigation, and selected the remedies accordingly; but at the expiration of one month, she had not improved at all, and then I insisted upon an examination, or, if she refused, intended to decline attending her. She consented, and an examination by means of the anal speculum at first gave me no insight into the difficulty; but with a fenestrated speculum, and by means of my fingers, I detected an erosion in the centre of a fissure, situated half an inch or more within the orifice of the anus. The sides or walls of this fissure felt thickened, granulated, and hypertrophied, and she experienced so much pain from the examination, that I gave her chloroform inhalations; but previous to this, she said: 'Now you touch the spot where I feel all my pains.'

"I immediately operated upon this fissure, using very carefully a sharp-pointed curved bistoury, making a simple incision through the mucous coat, so as to completely cut through the fissure to its bottom, and extending it longitudinally a little beyond its extremities.

"The patient was very irritable, hysterical, and unhappy, and had longed for offspring. The husband was perfectly healthy, and the only thing the wife had ever complained of was the infirmity in question. She always had distressing pains during defecation, and great suffering during coitus. These unpleasant symptoms all disappeared as soon as she recovered from the operation, which was within three weeks; and six months after, she became pregnant, and was safely delivered. The fissure of the anus was undoubtedly the cause of her sterility, although it had not been previously suspected. This ailment has been often overlooked by physicians."

FISSURE OF THE RECTUM.—In Dr. Baker Brown's classical work on *Surgical Diseases of Women* he devotes a chapter to "Diseases of the Rectum producing, or resulting from Uterine Disorder."

In this chapter he says there are three distinct kinds of fissure.

(1.) A small superficial ulceration just within the verge of the anus.

(2.) A deeper ulceration in the same position, which goes through the mucous membrane and into the muscular fibres of the external sphincter.

(3.) A true fissure or crack, not discoverable by external examination, as it is situated at least an inch from the anus.

Polypi usually accompany the two latter forms, and Dr. Brown thinks they are caused by the fissure.

The *first* form of disease, according to Brown, causes the most serious symptoms, referable to the bladder and uterus, namely, frequent desire to urinate, painful micturition, and incontinence of urine; also, dysmenorrhœa of a most painful character, vaginismus, a localized pain, almost constant, about six or eight inches up the bowels, great irritation about the anus and vulva; but rarely the painful defecation set down as a symptom of fissure. This last is a reason why the disease is often overlooked, and treatment applied to the uterus and bladder, instead of the rectum. This variety of fissure is often a cause of sterility, from the irritation which it keeps up in the uterus and vagina.

Treatment.—Dr. Brown's treatment, which I have adopted in several cases, is to draw a straight, blunt-pointed bistoury lightly through the ulceration and contiguous skin without the anus for about an inch. Curling,* in his admirable little work, which should be in the library of every physician, thinks that even in these superficial erosions, the muscular fibres "at the bottom of the sore," and for some distance each way, should be cut off.

In the *second* and *third* kinds a deeper cut, deeper than the fissure, through the fibres of the muscles and for an inch or two through the skin of the anus, is necessary.

These operations should be followed by a dressing of borated lint or cotton saturated with Glycerole of calendula. Before the operation the bowels should be thoroughly emptied by a laxative, or injections, and not allowed to move for a few days after.

In homœopathic practice some surprisingly favorable results have accrued from *internal* medication, where we are able to select the true *similimum*. In Volume V of the *North American Journal of Homœopathy* is an exhaustive paper by Dr. Perry on this subject, in which the indications for the following medicines are given, and several brilliant cures have been made by them:

These remedies are: (1.) Nitric acid and Ignatia. (2.) Plum-

---

* Diseases of the Rectum, p. 12.

bum, Arsenicum, Sulphur, Lachesis, Natr. m., Phosphorus, and Sepia. Dr. Roth adds Nux to the list, as a most important remedy. It seems, however, from some late experience, that *Ratania* ought to come after Nitric acid, as the next most important medicine. In the May No. (1878) of the *North American Journal of Homœopathy*, Dr. T. F. Allen gives a complete *résumé* of the powers of this drug, and reports a case where he made a brilliant cure with Ratania. He suggests Graphites, Pœonia, and Silica as good remedies. I have had good results follow the use of Æsculus and Aloes. Ratania, or any other indicated remedy, should be applied topically as well as given internally, for they have cured cases, when thus applied, when they were not given internally at all. Ratania is equally valuable for fissures or cracks in the nipple, vagina, or lips, showing that like Nitric acid it is generally useful, and that fissures are not always a purely local affection, but depend upon some constitutional dyscrasia.

Dr. Baker Brown believes that fissures, and other diseases of the rectum, especially irritability, spasm, and neuralgia, may be caused by the habit of "delection," or masturbation. In such instances he practices excision of the clitoris, as a means of cure, to be followed, if necessary, by the above surgical procedures on the rectum and anus. He shows conclusively that vaginismus is often caused by the same pernicious habit.

CHAPTER XII.

THE MECHANISM OF CONCEPTION.

A work on sterility would be incomplete if it did not mention certain important facts relating to the seminal fluid, and the processes which may be termed the "mechanics of conception."
This subject will be discussed under three heads, viz.:
  I. The nature and properties of semen.
  II. The method of its passage to the ovule.
  III. The successful conditions for such passage.

I. The Semen.

The seminal fluid is made up of the secretion of the testes, mixed with that of the seminal vesicles, prostate, and Cowper's glands. It is composed of the liquor seminis, granules, and spermatozoa.

Normal semen will drop from the end of a pipette or syringe in drops as easily as water. A small quantity falling into a glass of water is, by slight agitation, immediately diffused or dissolved in it. *Abnormal* semen full of mucus will not leave the mouth of the syringe quickly or suddenly, but ropes off for an inch or more before it breaks into a drop; but when it falls into water it preserves its tenacity, and but a small part of it is dissolved. It floats about in shreds, and eventually settles at the bottom of the glass in the form of a whitish sediment.

Semen destitute of spermatozoa has the usual sui generis odor, but lacks the appearance of uniformity that belongs to the normal secretion. When viewed by transmitted light, we usually see little whitish flakes of mucous floating through it. This is not always the case, however, for Sims mentions two instances in which it had the color and appearance of good semen, although wanting in spermatozoa. Abnormal semen is insoluble in hot or cold water, and floats about in it in cloudy flakes like ordinary mucus. It is more translucent than good semen, less milky, and less opaque. Under the microscope it presents the appearance of ordinary mucus. Good semen may, however, be loaded with mucus, which probably comes from the glandular apparatus at the neck of the bladder. The presence of a large amount of mucus is not necessarily inimical to the life of the spermatozoa.

The fructifying power of the semen lies in the spermatozoa alone, and not in the liquor seminis, or any fanciful "seminal aura."

Dr. Sims,* who had paid special attention to the examination of the seminal fluid, says: "If we take a drop of semen from the vagina immediately after sexual intercourse, and place it under the microscope, we shall see the hurried movements of seemingly thousands of spermatozoa. But this is not the best way of studying the phenomena of their movements.

Fig. 52.

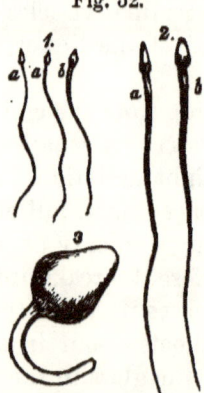

Spermatozoa (after Kölliker). 1. Magnified 250 diameters. 2. Magnified 600 diameters. *a.* Viewed from the side. *b.* Viewed from the back. 3. Fragment, head and one-fourth of tail. Magnified 2800 diameters (after Richardson).

The best plan is to take a drop of mucus from the canal of a perfectly normal cervix uteri some fifteen or twenty hours after

---

* Uterine Surgery, p. 315.

sexual intercourse. We shall then be better able to examine the spermatozoa; for we shall see them in the fluid that serves as the means of their finding their way towards the ovum.

"We shall find them moving more slowly, more continuously, if the term be allowed. Suppose we select any one spermatozoon for observation, and note particularly its various actions and movements. It will swim first one way and then another, or more in a straight line across the field of vision, and perhaps turn abruptly to retrace the path already traversed. If it encounters a large epithelial scale, it stops, places its head against it, as though trying to push it forwards, and when it fails to do so, it turns and moves off slowly in another direction, perhaps to encounter another opposing obstacle, to pause a moment and make another effort to overcome it, and then to turn again in search of some new field of exploration."

One would suppose, after reading this graphic description of the movements of the spermatozoa, that there should be no doubt about their possession of animal life, yet Carpenter, in his *Physiology*, writes: "Spermatozoa have no more claim to a distinct animal character than have the ciliated epithelia of mucous membrane, which likewise continue in movement when separated from the body. They appear to be nothing else than cell-germs, furnished with a peculiar power of movement, by which they are enabled to make their way into the situation where they may be received, cherished, and developed."

This statement would deny them all organization and life, as if *life* could be begotten *without* life. But let us inquire into the proofs of their possession of organization.

Locuenhoeck, Gerber, Valentine, Dujardin, Wagner, and other eminent microscopists, all testify to have discovered traces of *organization* in spermatozoa. Valentine says:* "The spermatozoa of the bear have a mouth, anus, and stomach, or a convoluted intestine." Commenting on this, Hassall † properly observes: "The determination of the fact that the sper-

---

* Muller's Embryology, with illustrations, p. 1475.
† Microscopical Anatomy, vol. i, p. 225.

matozoa are possessed of even the smallest amount of organization, would involve their classification in the animal kingdom." He further says, concerning their *motions*. "All the spermatozoa contained in a drop of semen which has undergone dilution, will not start into motion at once; many of them will remain for a time perfectly motionless, and then suddenly, as it were by an act of volition, begin to move themselves in all directions." Of their mode of progression, the same writer says: "The motions of the spermatozoa are effected principally by means of the tail, which is moved alternately from side to side, and during the progression the head is always in advance." It is said that the spermatozoa of different animals move in a different manner, because they differ very much in their form and structure. This would not be the case if they were "nothing more than ciliated epithelium."

Hassal also states that in the varied motions executed by the spermatozoa, they exhibit all the characters of volition; thus they move sometimes quickly, at others slowly, alter their course, stop altogether for a time, and then resume their eccentric movements. These movements it is impossible to explain by any reference to any hydroscopic properties which may be inherent in the spermatozoa, they appear to be so purely voluntary. Dr. Morris Wilson* says: "Spermatozoa when moving through a fluid *turn readily out of the way of any obstructions*, but they have not the backward movement of the vibriones." (Would ciliated epithelium avoid obstructions?)

Notwithstanding both Dalton and Draper in their works on *Human Physiology* assert that spermatozoa do not possess animal life nor organization, and affect to believe that the appearances above quoted were due to optical delusions, we feel confident that their distinctive animal nature will some time be fully proven. It will be more in accordance with true scientific modes of thought not to assert of spermatozoa a want of life and organization because we cannot now discover and demonstrate it, but await the result of more minute investigations.

Not only must there be living, moving spermatozoa in the

---

* Diseases of Seminal Vesicles.

semen, but their *motions must be normal and their life healthy*, or they cannot impregnate the ovum. Sims has shown that these organisms may be diseased* or crippled by the unhealthy medium in which they are ejected, or poisoned by the secretions of the uterus and vagina.

It is important to know under what circumstances the spermatozoa live longest and retain their healthy vitality. Under favorable circumstances they live many hours, even days: under unfavorable circumstances they die quickly. For instance, any great variation of temperature is fatal to their existence. In a temperature of 98° they will live a long time; a variation of ten degrees may cause death.

Certain conditions of the secretions of the vagina and uterus destroy their life. According to Donné, they live in pus and blood, but I imagine the pus or blood must be healthy. According to Wagner, they live in blood, milk, mucus, pus, syrup, and very delicate saline solutions. Sims says he has frequently seen conception happen when the cervix uteri was the seat of profuse suppuration. But if the secretions of the vagina be abnormally *acid*, it kills the spermatozoa instantly. The same may be said of the cervical mucus if it possess acrid or unhealthy properties, or if it is so thick and tenacious that they cannot move freely in it. (I have alluded to this when treating of *leucorrhœa*, etc.) "They do not exhibit any movements in pure semen, *i. e.*, before mixed with the fluids of the seminal vesicles" (Kolliker).

There are certain chemical and toxic agents that are poisonous to the spermatozoa, and others that are favorable to their vitality. Of the former are strong alkalies, acids, even when very weak, astringents, salt, alcohol, opium and all narcotic poisons, strychnia, and, indeed, all medicinal substances. They cannot live in urine and bile, or uterine mucus strongly alkaline (Donné). Hassall says the spermatozoa are devoid of life in persons who have died from the poisonous effects of prussic acid. This may be the case in poisoning with other diffusible poisons. In treating the causes of sterility we should bear

---

* Uterine Surgery, p. 352.

these facts in mind, and allow no medicinal substances to be injected or used in the vagina and uterus *near* the time of a coition which is desired to be fruitful.

To the last class, or those favorable to the life of the spermatozoa, are weak alkaline solutions, especially the phosphate of soda (Kolliker), or bicarbonate of soda (Sims). The strength should be about 5 grains to ℨj, and the water of the temperature of 98°. We may, therefore, use these alkalies when we wish to neutralize the too acid character of vaginal mucus. In healthy vaginal mucus spermatozoa may live several days, and in it their power of locomotion are certainly very extraordinary.

EXAMINATION OF VAGINAL AND CERVICAL MUCUS FOR SPERMATOZOA.—Dr. Sims[*] gives the following minute directions for procuring and examining the secretion of the vaginal and cervix uteri for spermatozoa: "Suppose we wish to examine the *vaginal* mucus soon after coition—say within an hour; we direct the patient to empty the bladder before the act, and to retain quietly the recumbent position after it. The dorsal decubitus is the best. To remove a few drops of the contents of the vagina, pass the index finger into it; press the posterior wall downward and backward, just under the cervix uteri; hold it so for a minute or two; the semen will necessarily gravitate to the pouch made by the pressure; then introduce the nozzle of the syringe along the finger; let it project slightly over the end of the finger-nail, and it will be easy enough to obtain what we want if there is any semen in the vagina. I am thus minute in explaining this simple operation, because we may fail in it entirely, even when the vagina contains large quantities of semen, if we neglect these minutiæ; and in this way, if we pass in the syringe in a haphazard manner and begin to draw the piston, the mucous membrane of the vagina is sucked up into the end of the tube, and thus it is possible to slide it around in various directions without getting a drop of mucus of any sort. But suppose we fail even with properly directed efforts; then the

---

[*] Uterine Surgery, p. 393.

left lateral position and my speculum will, in a moment, show us the whole contents of the vagina, and we can, with the syringe, remove what we want.

"When we wish to examine the *cervical* mucus, we should resort at once to the speculum and the proper position. It is well enough, then, to sponge away all the mucus from the vagina, and especially from about the cervix uteri, when we pass the nozzle of the syringe just within the os tincæ, and draw up a drop of its mucus. To do this, it is necessary to pull the cervix forwards, so as to be able to look into it, and to see exactly what we are doing. If the cervical mucus is very tenacious we may fail to get it away. Then it will at the next attempt be necessary, after introducing the syringe, and drawing up the mucus, to pass the left index finger to the edge of the os tincæ, and slide the end of the syringe on to the end of the finger without raising it from the surface of the cervix, or breaking its suction power. This may seem to be a little thing to describe so minutely, but really it is a most important matter to know and to do, if we expect to be exact in our investigations. The nicety of this manipulation renders it the more important for us to clear away all the vaginal mucus before we undertake it, lest we get some of this drawn up into the syringe, which would, of course, mar the precision of our observations. Suppose we succeed in this, then we may wish to pass the syringe up for an inch into the cervix to get a portion of mucus nearer the cavity of the uterus. This operation is quite as delicate and quite as important as the first, and is to be conducted in the same way. There is an object in having the end of the syringe bulb-shaped. This bulb fills up the os or the canal of the cervix, and prevents the air from being drawn into the instrument, as sometimes happened with me when it was slender and pointed."

A drop of this mucus procured from either vagina or cervix should be placed immediately under the microscope, for the spermatozoa will soon die.

In performing these operations or any other, there have been but two specula which could be used with convenience, viz., Sims's and Jackson's. Sims's instrument, or rather combination of instruments, requires an assistant in all cases where the operator has to use both hands. Jackson's was a great improvement on all other valvular specula, because the valves can be separated throughout their whole length.

I have, however, designed an instrument much lighter and more convenient than Jackson's. Through the kindness of E. H. Sargent & Co., surgical instrument makers, of Chicago, who manufactured the one I now use, I am enabled to present a cut of my speculum, of which the following is a description.*

The blades are concavo-convex; the upper blade is shorter than the lower, and falls into it when closed. The upper blade is slit, giving a long opening, through which the urethral orifice can be seen and any operation on the urethra or meatus performed. The upper blade, which comes under the pubes, is elevated by two methods—one, by applying the finger to the extension which projects upwards from its base, elevating the inner termination, as in Storer's and all other bivalves; the other, by means of a screw, which separates the upper blade

Fig. 53.

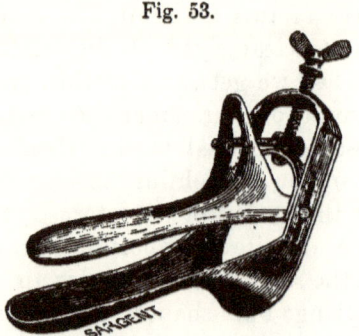

Hale's Expanding Speculum.

through its whole length from the lower or longer blade. This last gives it the quality of Sims's speculum, viz., a wide opening, through which almost any operation on the uterus can be performed without the aid of an assistant.

The small screw which is attached to the upper blade, and

---

* This speculum can be procured of Messrs. Boericke & Tafel, No. 145 Grand Street, New York, or ordered from any of their Branch Pharmacies.

projects through the lower arch, is used to fix the upper blade in position after its inner extremity has been elevated. In vaginismus, or small vaginas, the long screw which enlarges the opening between the two blades at their outer extremities can rarely be used without causing pain. But in cases where it is necessary to do so, it is much more powerful than Sims's, and can be left *in situ*, while both hands are otherwise engaged. This speculum can be introduced in the dorsal or lateral position. I generally use it with the woman on her back and with the lower extremities strongly flexed; although I frequently find it more convenient to use it in Sims's position, on the left side.

The foregoing excellent cut shows the mechanism of the instrument very plainly.

There should be a small hook projecting from the left side of the under side of the upper blade, to which a tenaculum can be attached, but it could not be represented in this cut.

II. The Passage of Spermatozoa to the Ovule.

Usually in women with healthy organs of generation there occurs during sexual congress, or sexual excitement, a profuse flow of thin, translucent mucus. This not only lubricates the vagina and inner surface of the vulva, but flows out upon the external parts. There are many facts recorded which prove unmistakably that if healthy semen be deposited in, and mixed with this mucus, even at some distance from the ostium vagina, the spermatozoa will, in a few moments, find their way through this secretion into the vagina and cervical canal to the cavity of the uterus. Sims records many cases which fully corroborate this assertion, and he goes on to refute the vulgar opinion, even now taught by some writers, that "to insure conception, sexual congress should be performed with a certain degree of completeness that would give an exhaustive satisfaction to both parties at the same moment." "How often," he says, "do we hear husbands complain of coldness on the part of wives, and attribute to this the failure to procreate; and sometimes wives are disposed to think, though they never complain, that the fault lies with the hasty ejaculation of the husband. Both are wrong."

The fact is that in either sex, neither passion or desire are necessary to conception. They are of no value except as *incentives*. "It matters not," says Dr. Sims, "how awkward and unsatisfactory the act of coition may be performed, so that the semen, with the proper fructifying principle, be placed in the vagina at the right moment; and, on the contrary, it matters not how perfectly and satisfactory it may be done if the semen lacks the fecundating power." My experience agrees with Sims, that both parties may possess intense passions and power of enjoyment and have no offspring; while in other cases be almost destitute of both, yet have a large family. Physicians, by teaching their patients the facts in this matter, may do much to render them happier in their domestic relations.

There are also some facts relating to conception which have a medico-legal importance. Does conception occur during intercourse, or soon after? Dr. Sims says: "If the uterus is in a normal condition we shall always, as a rule, find spermatozoa in the canal of the cervix *immediately* after coition." If in the canal of the cervix, why not in the uterine cavity? The spermatozoa certainly travel with considerable speed. Sims mentions one instance where a spermatozoa had travelled three and one-half inches from the surface of the hymen to the os tincæ in *three hours!* Conception or fecundation *may* take place immediately after, possibly during coition, if an ovum is in the cavity of the uterus ready to be impregnated; or it may be hours or days before impregnation occurs. Dr. S. R. Percy, of New York,* reports a case in which he found "living spermatozoa, and many dead ones," issuing from the os uteri eight and a half days after the last sexual connection. Sims thinks that spermatozoa never live over *twelve hours* in *vaginal* mucus; but in *cervical* mucus he has found living ones after *forty hours*. He knows of no reason why they may not live much longer. In some of the female mammalia they have been found in vaginal mucus eight days after copulation.

---

* Amer. Med. Times, 1861.

III. THE SUCCESSFUL CONDITIONS FOR SUCH PASSAGE.

It has been already stated that, after the semen has been deposited in the vagina, the spermatozoa find their way into the cervix, pass up into the uterus, and, if not finding the ovum, pass still further into the Fallopian tubes or ovary. Whenever and wherever they meet the ovum they enter it. *Contact* results in conception. Some microscopists have found the ovum transfixed by the spermatozoa, the head imbedded, the tail protruding and moving. In rare cases, the ovum after impregnation does not pass into the uterus, but remains where it was impregnated. Then occurs ovarian or tubal pregnancy. But the mechanism of their entrance into the cervix uteri is now a matter of considerable discussion. Gardner states (*Sterility*, p. 48), that "popular opinion declares that during copulation, and at the height of venereal sensation on the part of the female, the mouth of the womb *opens* to receive the semen, which passes into the cavity of the uterus." Since this was written, Dr. Sims writes: "I have over and over again examined the condition of the uterus after coition—often in four or five minutes after it—and I have usually found the following state of things: The uterus presents signs of exhaustion, if I may be allowed such an expression; for instance, if the uterus is in a normal position, or even moderately anteverted, we shall find the upper part of the vagina relaxed, and passively holding a large quantity of semen, in which the cervix uteri is submerged; the uterus itself seems to be fatigued, and drops by its gravity down towards the rectum, where it lazily sinks to the bottom of the little pool of semen." Dr. Sims believes that, in a normal condition of the uterus and cervix, the semen *suddenly* enters the cervical canal. His explanation of the process is: "The cervix is pressed forcibly against the glans (penis) by a contraction of the superior constrictor vaginæ; that this pressure necessarily forces out the contents of the canal of the cervix; that the parts subsequently become relaxed, the uterus returns suddenly to its normal condition, and the seminal fluid filling the vagina necessarily rushes into the canal of the cervix by a process similar to that by which a fluid would pass into an india-rubber bottle slightly compressed, so as to expel a portion of its contents before placing its mouth to a fluid of any sort." .... "From this it will be seen that I believe the cervix uteri to be shortened in the erethismal climax of coition by pressure exerted upon it in the direction of its long axis *when its position is normal*, which is impossible in a greatly abnormal position." This theory has an important bearing on the question of sterility, showing how it may be caused by *retroversion* or *anteversion*. For instance, in *retroversion*, "the fundus sinks still lower after coition than before, and this necessarily elevates the os tincæ still further from the seminal fluid, if any of it has been retained." In this case the spermatozoa, to enter the cervical canal, are obliged to "climb up," so to speak, on the cervix, and find their way into the canal. (It is in this condition that the lateral or "spoon-fashion" posture during coition may be advisable.) "In *anteversion*," says Sims, "we do not find the spermatozoa in the canal immediately after coition, because, with the os tincæ looking in the direction of the hollow of the sacrum, the same act and the same pressure would only force the anterior lip of the os tincæ up against the posterior lip, creating no vacuum and making no room for the newly introduced fluid." (In this instance the reversal of the natural position of the sexes during coition might remedy the defect.)

Dr. Montrose A. Pallen,[*] in a very elaborate and scientific discussion of this subject, says that during coition, "the posterior wall of the vagina is lengthened, and the engorgement of the vaginal plexuses further increases its elasticity and tension. When erotic excitement is at its height, the intra-pelvic erectility of the female is equally as tense as is the extra-pelvic erectility of the male organs, and the cavities of the neck and body of the uterus occupy a position midway between the axes of the inlet and the outlet, in a line corresponding to the centre of the upper portion of the vaginal insertion upon the cervix. Therefore, when the ejaculation of the semen takes

---

[*] The Philosophy of Menstruation, Conception, and Sterility (A Lecture, G. P. Putnam & Co).

place, it impinges upon the cavity of the cervix, and in many instances passes directly into the cavity of the uterine body. Many women contend that they can tell the very moment of impregnation from a peculiar sensation of sickening shock, not the excitability of orgasm, but a sickish, sinking pain, and the probabilities are that the germinating fluid is forcibly ejaculated to the very fundus, producing a decided sensation, which, if happening when there are none of the hyperæsthetic sensibilities of coition, would give rise to a very marked uterine shock or colic." All this may be true in women whose pelvic organs are in a normal condition and position. But how many women possess such organs? And when Dr. Pallen insists (p. 67) that conception must always be attended by this process, and warns us against believing that women are impregnated during sleep, intoxication, or when anæsthetized, he is going too far. His theory of "generative fixation," as he calls it, will not stand against the facts of common experience; for, without doubt, the majority of conceptions take place without any such process as he describes. Sims's theory is much more tenable. There is no doubt, however, that during sexual orgasms, the os generally opens or relaxes, and that more or less secretion of transparent mucus flows from the cervix, but that this is necessary to conception cannot be proved. Sims says the semen is "sucked into" the uterus *after* the uterine erection. Pallen declares it is propelled into the uterus during the "erectile hypertrophy." I do not believe either process is always necessary. If the cervix is filled with normal mucus, neither too thick, or too alkaline, or acrid, and the vaginal secretions are healthy, the spermatozoa are capable of "crawling from the posterior vaginal fornix over the infra-vaginal cervix into the uterine cavity," notwithstanding Dr. Pallen doubts their ability. There have been cases of undoubted *rape* when conception resulted, and to teach, as does Dr. Pallen, that conception without orgasm or "uterine fixation" was impossible, would be placing many good and virtuous women who had been violated, in a very painful position, and strip the law of all right to punish such an awful crime.

CHAPTER XIII.
HYGIENIC AND OTHER ERRORS CAUSING STERILITY.

There are many causes of sterility, in habits of life and conduct, which cannot be classified among the diseases mentioned in the previous pages. They are of such vital importance that to omit mention of them would greatly lessen the practical value of a work of this kind. It will not answer to pass over such errors of life and conduct in a cursory manner. Each one must be pointed out and described as minutely as professional delicacy will permit, but without that false modesty which is out of place in medical works.

One of the most injurious of all habits and customs to which the women of nearly all highly civilized nations are addicted, consists in improper and injurious arrangements of the clothing which they wear. This matter is so well written up by Dr. T. G. Thomas* that I take the liberty of quoting his entire section, entitled

IMPROPRIETIES IN DRESS.

"The dress adopted by the women of our times may be very graceful and becoming; it may possess the great advantages of developing the beauties of the figure and concealing its defects, but it certainly is conducive to the development of uterine diseases, and proves not merely a predisposing, but an exciting cause of them. For the proper performance of the function of respiration, an entire freedom of action should be given to the chest, and more especially is this needed at the base of the thorax, opposite the attachment of the important respiratory muscle, the diaphragm. The habit of contracting the body at the waist by tight clothing confines this part as if by splints; indeed it accomplishes just what the surgeon does who bandages

---

* Diseases of Women, p. 46.

the chest for a fractured rib, with the intent of limiting thoracic and substituting abdominal respiration.

"As the diaphragm, thus fettered, contracts, all lateral expansion being prevented, it presses the intestines upon the movable uterus, and forces this organ down upon the floor of the pelvis, or lays it across it. In addition to the force thus exerted, a number of pounds, say from five to ten, are bound around the contracted waist, and held up by the hips and the abdominal walls, which are rendered protuberant by the compression alluded to. The uterus is exposed to this downward pressure for fourteen hours out of every twenty-four; at stated intervals being still farther pressed upon by a distended stomach.

"In estimating the effects of direct pressure upon the position of the uterus, its extreme mobility must be constantly borne in mind. No more striking evidence of this can be cited than the fact, that in examining it by Sims's speculum, if the clothing be not loosened around the waist, the cervix is thrown so far back into the hollow of the sacrum as to make its engagement in the field of the instrument often very difficult, and that attention to this point in the arrangement of the patient will at once remove the difficulty. While the uterus is exposed by the speculum it will be found to ascend with every respiratory effort, and descend with every inspiration; and so distinct and constant are the rapid alterations of position thus induced, that in operations in the vaginal canal the surgeon can tell with great certainty how respiration is being affected by the anæsthetic employed. An organ so easily and decidedly influenced as to position by such slight causes must necessarily be affected by a constriction which, in autopsy, will sometimes be found to have left the impress of the ribs upon the liver, producing depressions corresponding to them.

"No one will charge me with drawing upon my imagination, even in the remotest degree, for the details of the following picture, for a little reflection will assure all of its correctness. A lady who has habitually dressed as already described, prepares for a ball by increasing all the evil influences which result from pressure. Although she may be menstruating, she dances until a late hour of the night, or rather an early hour of the morning. She then eats a hearty supper, passes out into the inclement night air, and rides a long distance to her home. This is repeated frequently during each season, until advancing age or the occurrence of disease puts an end to the process.

"A great deal of exposure is likewise entailed upon women by the uncovered state of the lower extremities. The body is covered, but under the skirts sweeps a chilling blast, and from the wet earth rises a moist vapor, that comes in contact with limbs encased in thin cotton cloth, which is entirely inadequate for their protection. It is not surprising that evil often results to a menstruating woman thus exposed.

"To a woman who has systematically displaced her uterus by years of imprudence, the act of sexual intercourse, which in one whose organs maintain a normal position, is a physiological process devoid of pathological results, becomes an absolute, a positive source of disease. The axis of the uterus is not identical with that of the vagina. While the latter has an axis coincident with that of the inferior strait, the former has one similar to that of the superior. This arrangement provides for the passage of the male organ below the cervix into the posterior cul-de-sac, the cervix thus escaping injury. But let the uterus be forced down as it is by the prevailing styles of fashionable dress, even to the distance of one inch, and the natural relation of the parts is altered. The cervix is directly injured, and thus a physiological process is insensibly merged into one productive of pathological results. How often do we see uterine disease occur just after matrimony, even where no excesses have been committed.

"It is not an excessive indulgence in coition which so often produces this result, but the indulgence to any degree on the part of a woman who has distorted the natural relations of the genital organs.

"But this is by no means the only method by which displacement of the uterus may induce disease of its structures. It disorders the circulation in the displaced organ, and produces passive congestion and its resulting hypertrophy, prevents the free escape of menstrual blood by pressing the os against the vagina, creates flexion, causes friction of the cervix against the floor of the pelvis, and stretches the uterine ligaments and destroys their power and efficiency.

"These facts should be carefully borne in mind by the physician who attempts to relieve uterine displacements by the use of pessaries. If he merely replaces the displaced organ, and relies for its support upon a pessary, he will often fail in accomplishing the desired result. He is striving at great disadvantage with a short lever power against the weight, not of the uterus alone, but of the superimposed viscera pressed downwards by several pounds of clothing, which add their weight at the same time that they constrict the waist, and substitute abdominal for thoracic respiration. Thus employed, the pessary will often give great pain, and so injure the parts upon which it rests as to necessitate removal, and the practitioner will find himself cast off from one of his most valuable resources. Should he, on the other hand, before employing a pessary, remove all constriction and weight from the abdominal walls, apply a well-fitting abdominal supporter over the hypogastrium, so as to aid the exhausted abdominal muscles in their work, keep the displaced and congested uterus out of the cavity of the pelvis by a tampon of medicated cotton, or bring gravitation to his assistance by the position of the patient, he will ordinarily at the end of a week be able to employ with great advantage the same pessary, which at first seemed to accomplish evil and not good."

Another subject mentioned by Dr. Thomas, and I have taken the liberty of quoting his remarks, is—

Imprudence during Menstruation.

This is also a very prolific source of uterine disease. Some women, through ignorance, many through recklessness, and a few from necessity, go out lightly clad in the most inclement weather during this period, and many suffer in consequence from violent congestive dysmenorrhœa, and often from endometritis. Every practitioner will meet with a certain number of cases of uterine disease which have this origin, and run on for years, ending, perhaps, in parenchymatous disease, which may prove incurable.

"During a period in which the ovaries and uterus are intensely engorged, in which the surface of the ovary is broken through by the escaping ovule, and the nervous system is in an unusual state of excitability, ordinary prudence would suggest that the body should be well covered, that the congested organs should be left at rest, and that exposure to cold and moisture should be sedulously avoided. I need not say that these rules are commonly neglected, and in evidence of the fact I will venture the assertion that, on this very day, the thermometer 15° above zero, the skating pond of our park contains scores of delicate and refined women, who are showing a disregard of them by their presence there.

"The immediate result of exposure during menstruation is most commonly inflammation of the mucous membrane of the uterus. Such an inflammation once excited will go on for years, and in time end in parenchymatous disease, entailing in its progress dysmenorrhœa, sterility, pelvic pain, and gastric disorders, which impair digestion and nutrition."

Passing on from these hygienic errors, I propose to mention certain errors of conduct as likely to prevent conception, namely:

EXCESSIVE VENERY AND ORGASM.—Too frequent indulgence in sexual intercourse, and excessive enjoyment of it, may both be causes of sterility. Dr. M. A. Pallen, in his *Clinical Lecture* heretofore quoted, says: "A very frequent cause of sterility is in excessive venery, producing a constant nerve stimulation of the generative circle, a constant congestion of the entire organism, which very decidedly interferes with normal and regular functioning. These causes are not infrequent in young married females, who failing to become impregnated during the earlier months of connubial life, are either taken to task by the husband or taunted by some more fortunate female friend, and as a consequence they repeatedly urge the act in the blended hope of offspring, and the fear of losing the respect and affection of their husbands. The result soon tells upon them both, constitutionally and sexually, and sterility is ingrafted upon a deteriorated physique, seriously complicating the treatment and many cases where the direction to remain after coition in the recumbent position till morning, resulted in conception after years of sterility. For this reason, if a fruitful coition is desired, the intercourse should take place just before the hour of sleep.

MEANS USED FOR THE PREVENTION OF CONCEPTION.—Among the most common of all the initial causes of sterility are the various methods adopted by men and women to prevent conception.

It is incredible that people united in marriage should deliberately set about preventing the most holy and natural result of love and the sexual union. I know well enough that there are cases where, from incurable diseases of the uterus, conception is uniformly followed by miscarriage and its attendant dangers. But these cases are so very rare that the dictum should lie entirely with the conscientious physician, and he alone should prescribe the method to prevent conception. So, also, are there rare cases of pelvic deformity where conception should be avoided; but, with all the improved methods of artificial delivery, the cases are rare indeed where such precautions are really necessary. As for those unnatural people who apply to physicians for prescriptions to prevent conception, because "it is not convenient," or because "their pecuniary circumstances are not suitable," or because they "wish to take a trip to Europe," I have no patience with or respect for them. No respectable or conscientious physician will aid and abet such criminality, for it is nearly as criminal to prevent a conception as it is to destroy the fruits of one, unless the above-mentioned irremovable physical obstacles render it necessary.

Among the worst means used for this purpose are:

(*a*.) The injection immediately after coition of *cold* water. This is especially injurious, because the vagina and uterus are flushed, congested, and bathed in mucus just after intercourse, and the cold douche acts as a sudden shock, contracting the full bloodvessels, suppressing the mucous discharges, and results in secondary abnormal congestion, and even inflammation. I have known many cases where the cold water was thrown directly into the uterus, through the open cervical canal, causing serious uterine colic, and in one case a dangerous attack of pelvic cellulitis.

(*b.*) Strong solutions of sulphate of zinc, plumbum aceticum, tannin, etc., are often used; and, even if used with *warm* water, the results are equally injurious. They set up in time congestion of the cervix, cervical leucorrhœa, etc., and the result is, that when these misguided people wish for children, they are surprised to find it impossible. They have brought on an acquired sterility very difficult of removal.

(*c.*) One of the most hurtful and abominable of all methods is the use of a so-called "pessarie" of French manufacture. (The French excel in such abominations.) It is a cup-shaped contrivance, nearly two inches in diameter, made of hard or soft rubber, which fits over the os uteri like a cap. It is not always effectual, but always injurious. One of the worst cases of chronic metritis with metrorrhagia in a beautiful and otherwise healthy woman, who applied to me for the cure of her sterile condition, was brought on by the persistent use for years of this infernal contrivance. Her husband had insisted upon the use of one made of *hard* rubber, and the delicate tissues of the cervix had been bruised until an incurable areolar hyperplasia had resulted.

Dr. Montrose A. Pallen, in his clinical lecture, says: "When the erection of the uterus subsides (after coition) the germinating fluid is imprisoned in the cavity of the body, and there retained. Hence the folly of many women who syringe themselves with cold water immediately after coition to prevent conception, unless the shock of the fluid begets some sudden contractile uterine action which expels the contained semen, a very rare occurrence indeed.

"The explanation why women who use cold water so rarely conceive, is not in the washing they give themselves, but from the fact that such an unnatural procedure had long since rendered themselves sterile by developing cervical catarrh in consequence of their having so frequently shocked the distended and engorged organs, thereby producing repeated blood stasis and temporary congestions.

ABORTION.—A well-known gynæcological writer says: "A single abortion from any cause, either from disease, or drugs, or instrumental interference, may induce sterility."

That this is true, the experience of every physician will substantiate. The pathological changes set up by the unnatural expulsion of the embryo at any period of gestation, are nearly always more or less persistent. After having one miscarriage, and this after a first conception, many women never become pregnant again. This occurs much oftener when instruments of any kind have been used to effect the purpose. Many women who have applied to me for the cure of their sterility, have tearfully confessed that a miscarriage caused by their own hands, during the first year of their married life, resulted in an entire inability to conceive. This result occurs when no examination which we are at present capable of making discloses any morbid condition of the womb. In these cases it is no stretch of the imagination to believe that the sterile condition was a direct divine punishment for the enormity of the sin committed.

I have known the most careful medicinal and topical treatment continued for years fail utterly in curing this condition with such antecedent cause. If this fact could be known and appreciated by young women, there would be much less of the crime committed; for, as a rule, women are almost entirely ignorant of the serious consequences of the act. Even when one or more miscarriages do not cause inability to conceive, they generally set up pathological changes in the uterus which result in *repeated abortions*, which in the end undermine the health of the victim, until complete sterility results.

Is it not time that, not only physicians, but clergymen and all public instructors, should boldly teach the lamentable consequence of this constantly increasing crime?

CHAPTER XIV.

GENERAL THERAPEUTICS OF STERILITY.

UNDER this head I propose to place those hygienic and therapeutic measures which find no special mention in the preceding pages. In examining a patient about to be treated for the sterile condition, the first object should be to ascertain the physiological, psychological, or pathological cause. If such cause be ascertained, the treatment will be found above; but cases will frequently occur in which the cause can not be satisfactorily ascertained. Either it is not appreciable to the senses, aided by the various aids to diagnosis, or it is not physical in its character. In either case, the physician must base his treatment on general principles, such as change of condition as regards climate, diet, exercise, etc., or he must follow the directions of Hahnemann, and select the remedy from the similarity of its symptoms with those presented by the patient, taking into special consideration the general sphere of action of the remedy (its *genius*), as well as its characteristic or key symptoms, and also its special or diagnostic symptoms. To further the proper selection of remedies in cases of sterility, the claims of certain medicines will be considered, and their characteristic indications pointed out. This cannot be done with absolute accuracy in this, any more than in other diseases or morbid states.

By referring to the list of *medicinal causes of sterility*, it will be seen that many of the same medicines are enumerated that appear below as the *curative agents in sterility*. That this should be so, is in accordance with *the* law of cure, which asserts that only those medicines which cause diseases are capable of curing similar ones. No medicine, therefore, can cure sterility without being capable, either directly or indirectly, of causing that condition.

This law is not confined to the action of medicinal agents, but extends to all others, such as climate, certain habits of life, mental conditions, and the like. That the curative influence of climate, particularly, is homœopathic in its action, is admirably demonstrated by Dr. T. K. Chambers.*

It is my belief that nearly all the medicines capable of causing sterility, do so by their action on the *ovaries*. Medicinal agents may, however, by their pathological action cause cervical leucorrhœa, displacements of the uterus, stenosis of the cervix, vaginismus, or the various other uterine and vaginal causes of sterility. But the function of ovulation is the central and all-important cause of reproduction. If this is interfered with, sterility results. Medicines which act pathologically on the ovaries, interfere with that function, and thus become sterility-causing agents.

The *ovarian* remedies for sterility are more especially:

Aurum.	Hamamelis.
Agnus castus.	Iodine.
Apis mel.	Iodide of lead.
Bromine.	Iodide of potassa.
Baryta carb.	Kali carb.
Capsicum.	Kali brom.
Caladium.	Lachesis.
Cannabis indica.	Phytolacca.
Conium.	Phosphorus.
Cantharis.	Platinum.
Chimaphila.	Ustilago.

The *uterine* remedies comprise a few of the *ovarian* and some others, namely:

Aurum.	Helonias.
Aletris.	Lilium.
Borax.	Mercurius.
Cantharis.	Moschus.
Cimicifuga.	Phytolacca.
Caulophyllum.	Ruta grav.
Eupatorium purp.	Sabina.
Gossypium.	Secale.

---

* See Monthly Hom. Review, March, 1868.

Sumbul.
Stillingia.
Sepia.

Senecio.
Ustilago.
Viscum album.

CHARACTERISTIC INDICATIONS.—It is impossible in a work of this size to give the general and special symptoms, as well as the pathological indications for the use of the above enumerated medicines. Many of them have been mentioned in the preceding pages in connection with the diseases causing sterility. I have, therefore, thought it best to give only the "characteristics" of the remedies as a guide to their selection.

Aurum.—This remedy is *secondarily* indicated in *amenorrhœa* dependent on torpor of the ovaries, in scanty menstruation with chronic metritis; in sterility dependent on these states, or due to "coldness," or female impotency with suicidal depression. (Doses, a few grains of the 2d or 3d trit.)

Gold is *primarily* indicated for symptoms similar to platinum, namely: profuse and frequent menses, congestion of the uterus, increased sexual desire, and mental or emotional irritability. (Dose, the 12th to 30th.) I prefer the Aurum mur. or the Muriate of gold and sodium.

Agnus castus.—A complete loss of sexual power and desire; albuminous leucorrhœa; ovarian atony, or suspension of the function of ovulation; amenorrhœa; melancholy.

Aletris.—General debility; paleness; inability of the uterus to retain the product of conception; sterility after abortion.

Asarum (*Can.* and *Europ.*)—Said to cause sterility.

Apis mel.—Chronic or acute ovaritis with *stinging* pains, followed by ovarian enlargement. In women who are troubled with urticaria and pruritus pudenda, also dysuria, with stinging pains during urination.

Borax.—"A female had been sterile for fourteen years, on account of a chronic acrid leucorrhœa; she received, among other remedies, *Borax*, after which she became pregnant, and the leucorrhœa improved. *Easy conception* during the use of *Borax* observed in five women."—(*Hahnemann's Chronic Diseases*, II.)

It is indicated when membranous dysmenorrhœa causes

sterility; also for abrasion or erosion of the os uteri, and aphthous affections of the vagina and cervix.

Baryta carb.—Loss of sexual desire and power; very scanty menses; general appearance of premature old age; catches cold very easily. (Also Baryta mur. and iod.)

Cantharis.—Sterility with great sexual excitement, with irritation of the ovaries (primary effect); or sterility from the opposite condition, viz., loss of ovarian function (secondary effect).* For the first-named condition give the 30th, for the latter, the 1st. It has been used successfully by allopaths. I have known several cases cured by doses of 5 drops of the tincture, three times a day; it caused violent dysuria, but conception resulted.

Capsicum.—For fat women who are generally chilly, have a masculine appearance; atrophy of the ovaries (?) with amaurosis.

Calcarea.—Sterility in adipose women, who have profuse and too frequent menses.†

Caladium.—Sterility with melancholy, faint feeling in stomach, fetid urine, asthma, loss of sexual desire, coldness and cold sweat of the sexual organs.

Cannabis.—(Cannabis ind. is preferable.) "Great excitation of the sexual instinct, accompanied by sterility." Profuse menstruation, with various urinary difficulties.‡

Chimaphila.—Atrophy of the ovaries (?) and mammæ. The urine is full of mucus; scaly eruptions on the skin.

Conium.—One of the chief remedies in sterility. Atrophy of the ovaries and mammæ; albuminous or acrid leucorrhœa; loss of sexual power and desire; amenorrhœa, or very scanty menses; pain and swelling of the breasts before the menses.

Professor Tully (allopath) says:§ "In serveral cases, where

---

* Gardner on Sterility, p. 159.
† "*Calc.* (x) given against a copper eruption in the face, effected pregnancy in a woman who had not conceived for nine years past."—*Ruckert's Therapeutics*.
‡ "*Cannabis* (1st) one drop, and afterwards Merc. (3d). Giving both remedies twice to husband and wife, effected pregnancy, after a sterility of six years."—*Ruckert's Therapeutics*.
§ Materia Medica.

persons had been married eight or ten years without offspring, I have known a thorough use of good extract of *conium* employed by the party with whom the difficulty was supposed to exist, result in the subsequent birth of one, two, or more children. Again, when, in the course of a few years after marriage, there had been one, two, or more children, and then a suspension of childbearing for ten or a dozen years, after a thorough course of good extract of *conium*, taken by the party in fault, I have known one, two, or more children born. I recollect prescribing extract of *conium* for a lady who had ceased for a long time to bear children. At the time of my prescribing for her, she had some chronic difficulty, I forget what it was, though it could not have been of long duration. I made several calls, at comparatively long intervals, and at each was informed that she was better, and apparently continuing to improve. At last, at one of my visits, her conduct was very petulant, to say the least. She told me she had discontinued the medicine, and should take no more of it; but I could not ascertain the reason of her rejection and refusal of it. In the mean time her husband sat by, laughing. At last the lady left the room, apparently quite angry. Her husband then informed me that she had just become gravid, and at a time when she considered that her previous children were too old, and at what she considered too late a period of her life for childbearing, since she was a little past forty years of age. He said she considered her existing condition as due to my medicine. I answered that I thought she must, at least, divide the blame between him and my medicine." (Tincture of Conium should be made from the *unripe* seeds.)

Cimicifuga.—Sterility with hysteria, spinal irritation, want of vitality in the ovaries and uterus (congested cervix). She is subject to melancholy, severe congestive headaches, chorea, and acute rheumatism. (I have known many cases of sterility cured by *Cimicifuga* when all other means failed.)

Caulophyllum.—Sterility with spasmodic dysmenorrhœa; she is subject to rheumatism of the fingers and hands.

Eupatorium purp.—Sterility in women who suffer from nervous exhaustion, have profuse flow of watery urine from the least excitement; loss of sexual desire, uterine and ovarian atony; sterility from frequent abortions.*

Gossypium.—May prove of value in some cases of sterility. Supposed to be useful in uterine atony from frequent miscarriages.†

Helonias.—Sterility, with chlorosis, debility, diabetes, atony of the generative organs, female impotence, prolapsus uteri, melancholy, anæmia.

Iodine.—Sterility, with emaciation, weakness, especially in the knees; atrophy of the ovaries and mammæ; goitre. The testimony in favor of *Iodine* in sterility from allopathic sources, is worthy of our attention. Professor Tully says:‡ "A free and protracted use of *Iodine* is said to prove aphrodisiac, *i. e.*, to increase venereal appetite, and the power of gratifying it. Valetudinarian women who have been married a number of years without children, not infrequently become gravid after a thorough course of *Iodine.*"

In the latest pathogenesis§ the following recorded symptoms are found:

"A case is said to have occurred where the female became barren soon after commencing the use of Iodine. Before she commenced the use of Iodine, she gave birth to a child annually, but from the time of commencing its use to the present, a period of eight years, she has never become pregnant."

(These symptoms were quoted from Dr. Rivers, *American Journal of Medical Sciences*, 1831, p. 546.)

No clue is given by Dr. R., to the pathological conditions of the sexual organs which it caused. My conviction is, that the sterility was due to arrest of ovulation, but it may have been uterine catarrh.

Bromine has an effect on the ovaries similar to Iodine.

Iodide of Lead.—In atrophic ovarian sterility. (See *N. A. Journal of Hom.*, vol. ii.)

Iodide of Potash.—(See Iodine.)

---

* New Remedies, p. 369.
† Ib., pp. 475, 539.
‡ Materia Medica, p. 2.
§ Encyclopædia of Materia Medica, vol. v.

Kali carb.—A prominent medical author asserts that the sterility of American women is, to a great extent, due to the excessive use of *carbonate of potassa* in bread. (See *Hahnemann's Chronic Diseases.*)

The Bromides.—All the bromides cause extinction of sexual desire, and a general paralysis of the reproductive organs. They are primarily indicated in sterility from deficient ovulation and sexual inactivity; and, secondarily, in excessive excitation of the sexual organs, erotomania, etc.

Lachesis.—Chronic inflammation and other destructive diseases of the ovaries, also many diseases of the uterus, leading to sterility.

Lilium tig. will doubtless prove a valuable remedy. (See pathogenesis in *New Remedies*, vol. i, or *Allen's Encyclopedia*).

Mercurius.—It has been observed that women under the action of *Mercury* easily conceive.* It has the power of causing sterility, as all authors attest. It cures sterility from syphilis; also sterility from poisonous leucorrhœa.

Moschus.—This odorous drug is but little used in homœopathic practice, although it appears to have a sphere of action nearly equal to many of our much-used agents. It is a nearer analogue of *Platinum* than many suppose. Hahnemann says: "It compares with *Asafœtida, Crocus, Coffea, Conium*, and *Stramonium.* Its primary action on the generative organs is decidedly excitant, like *Phosphorus, Cantharis*, and *Platinum.*" The following symptoms recorded by Hahnemann are proof:

"Violent sexual desire, with intolerable titillation; drawing and pressing towards the sexual organs, as if the menses would appear; menses too early by six days, and too profuse. Menses appear even when merely smelling of the drug; reappearance of menses, which had stopped a whole year; violent drawing pains at the appearance of the menses."

This violent excitement of the ovaries and uterus must necessarily be followed by impotence and *sterility*, from exhausted vitality. Many lymphatic, obese, and hysterical women present all the above symptoms of its primary action, and are also

---

* Symptomen Codex.

sterile. In such cases *Moschus* in the 6th trituration ought to remove the condition. If the secondary effects obtain, give the lower triturations.

Mangeti, writing in 1703, thus descants of the virtues of *Moschus* in female sterility: "There may be various causes of hindered conception, but that they lie very frequently in an excess of cold or wet of the uterus, or an obstruction of the uterine vessels and ducts is established; which fact Hippocrates (sec. iv, Aphor. 62) expresses in these words: 'They who have cold and crowded (down) wombs, do not conceive. And they who have wet wombs do not conceive; for the germ is extinguished in them.'

This choked state, Lud. (*i. e.* Louis) Mercatus (book iii, chap. iii, on *Affections of Women*) thinks is the result of abundance of thick or cold humors. But that wombs that have been more hurried than was suited to their nature should be dried, and that fumigation should be applied to them. Hippocrates has shown (book i, on *Affections of Women*): "From these considerations it is concluded that Musk can relieve this evil, since it warms too cold blood and serum, dissolves those that are too thick, properly dries off too hurried parts, thins phlegm and mucus, and thus opens obstructions, not only of the vessels, but also of the passages and canals. Thus the virile semen may be not molested and weakened, nor prevented from extreme penetration, but nourished by moderate warmth, may be borne away from impediments to the female ovary, and thus the eggs, being fecundated, may in due time receive free transit into temperate wombs. To this end, Dom. Leonus (in *Medical Art*, book vi, chap. xiii, sec. 3), prescribes an electuary (confection), as tried on many, which is compounded with musk."

"John Bapt. Porta tells (*Mag. Natur*, book viii, chap. vii,), that he has found that a fresh-laid egg, with which a single grain of *Musk* has been mixed, is, when swallowed (absorptum) excellent to assist conception. A fumigation (or perfume) and a pessary with musk, are described in Roderick Castentris, book iii, sec. 1, chap. ii."

It will be observed that the old physicians advised it for the symptoms resembling its secondary effects. There may be more hidden under this absolete language than at first seems. *Sumbul*, or *Castoreum*, may be equally useful, as they are nearly allied to Musk.

Morphine.—Dr. Allen quotes (*Encyc. Mat. Med.*) Lerinstein (*General Effects of Morphine*), who says, in relation to effects on the female generative organs: "The amenorrhœa of Morphine is gradually developed from dysmenorrhœa, or it occurs suddenly. Conception has never been noticed in (these) amenorrhœic women, who had been repeatedly pregnant previous to the use of Morphine. Therefore, it seems probable that the cessation of the menses is dependent upon anomalies of the ovaries. According to Pflugus's theory, in the amenorrhœa of Morphine the growth of the cells of the ovaries cease from one period to another, and, in consequence, the irritability that is transmitted by the ovarian cells, and causes, on the one hand, bursting of the Graafian follicles, and, on the other, determines the reflex condition of congestion of the sexual organs, is wanting. As a consequence, Morphine affects the ovaries as it does other secretory glands, namely: it renders them unable to perform their functions. It is most probable that menstruation ceases because ovulation ceases. The supposition that Morphine injections cause the arrest of the functions of the organs of generation is justified by the fact, that after the cessation of Morphine these organs recover their activity. The sexual desire is at first increased by the habitual use of Morphine, but after the graver symptoms of poisoning have been developed it almost entirely disappears.

"It is also noteworthy that women who have suffered from fluor albus, are generally free from it during the prolonged use of Morphine; this reappears after the withdrawal of the drug, often with labor-like pains."

The same author says Morphine causes impotency in men.

It would seem by this testimony that Morphine ought to be a specific for some forms of sterility, namely: where it has been preceded by undue sexual erethism (passion, erotism), and where the excessive indulgence has weakened the virile functions of the ovaries especially, and amenorrhœa occurs from "lapse of the menstrual habit." In a late pamphlet by Dr.

GENERAL THERAPEUTICS OF STERILITY.   271

Montrose A. Pallen, he contends that menstruation depends not alone on ovulation, viz., "that menstruation is a neurosis, indicating anatomical changes, hyperplastic action, degeneration of tissue and reparation process;" but he does not admit that menstruation ever can occur in congenital absence of the ovaries, while "a bleeding, but not menstruation" may occur in cases of *extirpation* of the ovaries. (See Jackson, *On Ovulation*.)

Morphine evidently does not cause atrophy of the ovaries or paresis, but simply *arrest of function*, which returns when the drug is suspended. When, therefore, we meet with cases in practice where we believe this condition exists in amenorrhœic women, or women with very scanty menses and little or no sexual desire, it seems probable that small doses of Morphine (2^x or 3^x) might aid in restoring the functions of ovulation and the ability to conceive.

Phytolacca may be useful in sterility from syphilis, or from atrophy of the ovaries; uterine leucorrhœa, acrid, with ulceration or erosion of the os.

Phosphorus.—This remedy has probably a greater reputation for the removal of sterility than any other. It is used successfully by both schools of medicine—by allopathists in large doses, by the homœopathist in minute doses—cures following its administration, whether used in massive or infinitesimal quantities. There must be some rational explanation of this, else homœopathy is not the only true system of therapeutics.

The action of *Phosphorus* on the healthy organs of generation consists of two series of effects, namely: a *primary action*, during which it stimulates the ovaries to excessive functional irritation, which often proceeds to hyperæmia and inflammation. This ovarian excitation extends to the uterus, vagina, and clitoris, resulting in hyperæmia and hyperæsthesia, with inordinate sexual desire, uterine spasms, albuminous leucorrhœa, menorrhagia, ulceration, and various other morbid states; all causes of sterility. After the primary irritation has exhausted the vitality of these organs, the *secondary effects* set in, marked by loss of ovarian function (deficiency or suspension of ovulation), various organic diseases of the ovary, such as atrophy, tumors, etc., and a consequent loss of tonicity in the uterus and other generative organs, resulting in loss of sexual feeling, scanty menses, too pale and delaying, acrid, excoriating acid leucorrhœa, and many others of the various conditions leading to sterility.

Phosphorus is, therefore, homœopathic to sterility by virtue of its primary and secondary effects. Material doses will cure sterility if the secondary symptoms are present. It is for these symptoms and conditions that it is used successfully in allopathic practice. In the middle and highest potencies it will cure sterility presenting symptoms similar to those enumerated among its primary actions. Gardner* quotes Dr. Mackenzie, who, before the Medical Society of London, "recommended *Amorphous phosphorus* in certain affections of the uterine organs, attended with weakness and irritability of the nervous system, to be given in doses from 10 to 30 grains, diffused in water. It appears to act as a direct tonic or stimulant to the uterine system." He has known pregnancy to supervene upon its employment, after a lengthened period of sterility subsequent to marriage.

It is more relied upon than any other remedy in the actual impotence of men, *i. e.*, from loss of power in the testicles to secrete semen. Any remedy that will restore this power in men, will restore the function of ovulation in sterile women.

In Ruckert's *Therapeutics*, we find the observation that "*Phosphorus* appears to be a distinguished remedy against sterility."

Several of my colleagues have informed me that they have known conception to occur in women supposed to be sterile, while under the administration of *Phosphorus* in minute doses.

I can recall several instances occurring in my practice, when women who were taking this remedy continuously for pulmonary disorders, unexpectedly became pregnant for the first time in many years.

Phosphorus very closely simulates *Cantharides* in its action on the reproductive organs.

---

* On Sterility, p. 160.

Phosphoric acid or the Hypophosphites may be used instead of Phosphorus in some cases.

Platinum.—This metal appears to act in somewhat a similar manner to *Phosphorus*, but its characteristic symptoms differ from that remedy. Thus the sexual excitement of *Platinum* is attended by melancholy, hysterical symptoms, cramps, with coldness and torpor, while that of *Phosphorus* is marked by joyous excitement, febrile heat, and nervous erethism. Gollman* has *Platinum* indicated "when the patient feels oppressed with anxiety, or feels very cheerful one day and low-spirited the next."

It is indicated in subacute ovaritis, with pain as if beaten, congestion of the uterus, induration of the cervix, menorrhagia, and albuminous leucorrhœa. (I prefer Plat. *mur*. *Murex* is its nearest analogue.)

Pulsatilla.—Both species of *Pulsatilla* act on the organs of generation of women in such a manner as to cause sterility. Ovarian irritation, uterine, cervical, and vaginal leucorrhœa, uterine congestions, etc., are the conditions generally cured by them. For characteristic symptoms, consult their pathogeneses. (Clematis is very closely allied to Puls.)

Ruta grav. causes abortion, followed by too frequent, scanty menses, with corrosive leucorrhœa and sterility.

Sabina.—This drug acts as a tissue irritant to the ovaries and uterus; the latter, in particular, causing primarily intense active congestion, inflammation, and disorganization; and, secondarily, passive congestion or anæmia, with torpor and actual paresis. It is, therefore, homœopathic to sterility, with excessive uterine irritation, frequent and profuse menses, acrid leucorrhœa, ulceration, as well as sterility from uterine torpor, chronic congestion, ulceration, amenorrhœa; very scanty menses or passive metrorrhagia. It is analogous in action to *Erigeron, Tanacetum, Trillium, Crocus, Senecio,* and *Calcarea.*

Sanguinaria and *Stillingia* both have aphrodisiac properties, and may prove useful remedies in the conditions which lead

---

* Disease of Sexual Organs, p. 251.

to sterility. Sanguinaria resembles Calcarea, while Phosphorus and Stillingia resemble Phytolacca and Iodine.

Secale.—Ergot has a different sphere of action from *Sabina*. It is a uterine motor stimulant. In its primary action it can excite the uterus to such spasmodic irritation that the ovule, before or after impregnation, is thrown from the womb or destroyed in its cavity. In its secondary action it causes an actual uterine *paresis*, in which it is impossible for the impregnated ovum to be retained or nourished. It will cure sterility when prescribed in the medium or the lowest attenuations, according as the symptoms resemble its primary or secondary action. Guernsey says it is always indicated in "thin, scrawny women, with low vitality, subject to hæmorrhage of thin, black blood, aggravated by the least motion." This is a part of its secondary action, however.

Sepia.—This agent, by its profound disturbing action on the generative organs, is capable of causing sterility resulting from a variety of diseased conditions. It is best indicated in torpid, chronic stages of congestion, ulceration, and degeneration of the uterine tissues.

For a correct appreciation of the genius of this remedy, the physician should study its pathogenesis and the observations of our best authors.

Stillingia will remove sterility resulting from secondary syphilis or chronic mercurialization. Its action in this respect compares with *Aurum, Phytolacca, Iodide of potassium,* and *Corydalis.*

Senecio.—This new remedy is not sufficiently appreciated by physicians. It is as important as *Pulsatilla, Calcarea,* and *Sepia* in the treatment of the diseases of women. The conditions indicating its use in sterility are: Frequent, painful, and profuse menses, cervical leucorrhœa, irritable uterus, subacute ovaritis (primary effects), or delaying and scanty menses, dysmenorrhœa, with scanty flow, ovarian torpor, chronic mucous leucorrhœa (secondary effects). In these conditions it will rarely disappoint us if we prescribe it in the proper dose and persist in its use.

Ustilago.—This remedy is likely to prove of value in the

treatment of sterility, as well as miscarriage. It appears to be a near analogue of *Secale, Caulophyllin,* and *Cimicifuga,* but causes a more profound degeneration of the tissues and nervous system, exceeding even *Secale.* The symptoms indicating its use may be found in *New Remedies,* vol. i.

In addition to the above remedies the following are recommended in the treatment of sterility. Gollman mentions *Dulcamara, Sulphur, Cocculus, Ignatia, Arsenicum, China, Nux vomica.*

Jahr refers to *Borax, Amm., Caust., Graph., Natrum, Sulph. ac., Cicuta, Croc., Dulc., Ferr., Hyos., Natr.,* and also those already mentioned.

Hempel recommends *Canth., Ferr., Plumbum,* and *Sabina.*

Dr. Pulte, of Cincinnati, contributes his experience in the use of the following remedies. "The remedies mostly indicated in cases of sterility coming under my observation, have generally been those known best under the name of uterine remedies, such as *Belladonna, Sepia, Phosphorus, Calcarea, Carb., Sulphur,* etc. Of these I remember to have used oftener *Sepia* than any other remedy, simply for this reason: that the symptoms present indicated more frequently its use. As the causes of sterility are so various and many, it is evident that a uniform treatment for its cure cannot be established. Yet I have seen sterility disappear where the lady had not been treated for that special purpose, but for other ailments, which, however, were connected with it, as the result showed. It is to be hoped that this subject will continue to engage the earnest attention of the profession, as a more frequent removal of sterility will counterbalance the evil effects of artificially but criminally instituted abortion; it will be the means to bring happiness in households who deserve to bear children.

MISCELLANEOUS THERAPEUTICAL AGENTS.—*Galvanism* or *electricity* has obtained considerable reputation for the removal of *sterility.* M. Rouband, in his recent work, states that he "has had greater success than most in the treatment of cases that I am now describing when no disease is apparent." Indeed, he claims that, "in six out of ten cases, particularly in those whose sterility is from sexual abuse or prostitution, this treatment is followed by childbirth. The application is made once or twice a week for several months, during which time coition is suspended, and great simplicity and regularity in all the details of life are to be rigorously exacted."*

Gollman recommends galvanism, and advises that it be applied by placing one pole on the os uteri, the other on the sacrum, or at various places along the spine.

In Braithwaite we find recommended,† in cases of defective ovarian action, an application of Professor Recamier's galvanic poultices.

I have heard of a few cases where the so-called electro-galvanic baths appeared to have resulted in the removal of the sterile condition.

CARBONIC ACID GAS.—The topical application of this gas, according to the method of Dr. James Johnson,‡ is said to have exerted a remarkably powerful influence upon the reproductive organs. "Females who had for years sighed in vain for pregnancy, became fertile in a very short space of time after this gas had been applied in douches to the uterus, by means of a pipe introduced into contact with that organ through the vagina. Cases in which this beneficial result had followed its use were so numerous, that the subject merited consideration."

GYMNASTICS.—The systematic adoption of gymnastic exercises, and especially the "Swedish Movement Cure," or "new gymnastics," will doubtless be of great value as an aid to the physician in his treatment of sterility, when not caused by any organic obstruction.

HYDROPATHY, or the systematic application of water, has doubtless resulted in the removal of the sterile condition. As an adjunct to homœopathic treatment, the water-cure processes may be advised.

---

* Gardner on Sterility.
† Part xxiv, p. 348.
‡ Braithwaite, part ii, p. 64.

MAMMARY IRRITATION.—Dr. Marshall Hall throws out a suggestion on the treatment of *sterility*, which may be tried under some circumstances. "There is an extraordinary sympathy between the mammæ and the uterus, so that a functional condition of the former influences that of the latter. This sympathy is partly nervous in its character, partly vascular. As a reflex action, the uterus is made to contract after parturition, by applying the newborn infant to the mammæ; as a vascular sympathy, uterine hæmorrhage and leucorrhœa occur from undue lactation."

In cases of sterility arising from a temporary suspension of function, it is a question whether or not the uterus can be stimulated so as to assume a healthy functional action in the way suggested by Dr. Hall, who says: "My suggestion, then, is, that when the mammæ are excited at the return of the catamenial period, a robust infant be repeatedly and perseveringly applied, in the hope that the secretion of milk may be excited, and that the uterine blood may be diverted from the uterus and directed into the mammary vessels, and that a change in the uterine system and a proneness to conception may be induced. I would propose that the patient should sleep for one week before, and during the catamenial period, with an infant on her bosom." (Braithwaite, part ix, p. 202.)

Remarks.—It is possible that the irritation of the mammæ, as above recommended, would so change the condition of the uterus as to cure sterility. In such cases, however, other means of irritation might be tried, as an infant is not always "convenient." Poultices or sinapisms might be applied, or what would be better, an application of the leaves of the Ricinus communis, or castor-oil plant. Dr. McWilliams first published an account of the remarkable properties of this plant, which, when applied to the breasts of virgins even, will produce a profuse and natural flow of milk, sufficient to enable them to act as wet-nurses. Dr. Tyler Smith bears testimony to the efficacy of the leaves to produce this effect. The whole account of the method of applying the leaves will be found in the *Lancet*, 1850, p. 294, or Peter's *Diseases of Married Women*, p. 129.

INJECTION OF SEMEN INTO THE UTERUS.—This method of removing sterility is fully described by Dr. Sims in his work on uterine surgery. It has been recommended by several medical authorities, some of them very old. Dr. Sims thinks it may be successful in some cases, and he gives a few instances in which he tested the plan in his own practice, with the desired result. When all other means fail, and the desire for an heir is above all other considerations, this method can be advised by the physician.

THE TREATMENT

OF

GENERAL AND SPECIAL

DISORDERS AND ACCIDENTS

OF

PREGNANCY;

ALSO OF

PAINFUL AND DIFFICULT LABOR.

PREFACE.

THE two chapters on Dystocia which I contributed to Dr. Richardson's "*System of Obstetrics,*" were so favorably received, and awakened such an interest in the subject, that I have rewritten them and made copious additions.

There are many reasons why the profession should pay greater attention to the subject of Dystocia. They should disabuse their own minds and that of the public of the false idea that the sufferings of pregnant women must be borne without relief.

When women become convinced that the sufferings of gestation and the pains and anxieties of labor can be alleviated, great good will be accomplished.

This belief will remove many of the incentives which lead women to the crime of fœticide. It is the fear of suffering and death which influences most women to desire to avoid pregnancy, and thus to violate the purest instincts of maternity.

Free a woman's mind from such fear and anxieties, and a favorable influence will be exerted by the mind of the mother upon the spiritual and mental nature of the child.

If anything I have written in the following pages shall prove of benefit to womankind, making her more happy in her maternal relations, I shall feel amply repaid for all my investigations and labor.

E. M. HALE.

CHICAGO, 1878.

DYSTOCIA.

CHAPTER I.
GENERAL DISORDERS OF PREGNANCY.

By Functional Dystocia is meant those forms of preternatural and painful labors which are due to abnormal conditions of the generative organs, *not* structural in character.

The *causes* of Functional Dystocia may be divided into three classes. Either class may interfere with the regular processes of nature.

1. Those special disorders and accidents occurring during pregnancy; also, those *accidents* which may occur *during* labor, all of which may endanger the life or health of the mother or her child.

2. Those labors rendered difficult, impossible or dangerous, by a *deficient* or *excessive* action of the expulsive forces.

3. Those rendered difficult, impossible, or dangerous, by obstacles (*non*-structural) to the expulsion of the fœtus.

The *treatment* of Functional Dystocia may be divided into:

1. *Preventive*, which includes those hygienic measures and medicinal agents which may be advised during pregnancy and previous to the time of labor.

2. *Immediate*, or those medicinal agents and auxiliary measures to be adopted during the progress of the labor.

PREVENTIVE TREATMENT.—The adoption of a prophylactic treatment for painful or difficult labor has not attracted that attention from physicians that the subject demands.

It is a singular and suggestive fact that nearly all barbarous and semi-civilized tribes or nations *do* adopt some treatment to which pregnant women are subjected, and which is supposed to have a modifying influence on their confinements.

This treatment, in some cases, consists merely of certain baths and ablutions; in others, of semi-religious or superstitious rites, such as ablutions and alterations in diet; while in some tribes, like the North American Indians, the treatment consists almost wholly in the drinking, for several months, weeks or days previous to labor, of the infusions or decoctions of indigenous plants.

Some physicians affect to sneer at any medicinal measures used to ward off possible abnormal conditions which may render labor painful and difficult. But if these same physicians could *know* that any patron would, after a few weeks or days, be attacked by scarlet fever or rheumatism, would they not prescribe some hygienic or medicinal measures to mitigate the severity of such attacks?

Now, although labor is a natural function, and the resources of the organism are usually sufficient for its accomplishment, yet there are a number of circumstances which may interfere with the work of nature, and render the process painful, difficult, and dangerous.

If, then, the physician can devise any means by which these obstacles to a normal labor can be prevented or modified, it is his duty to adopt such means for the relief of those who are placed under his care.

I do not believe that it is necessary, or normal, that *all* pregnancies or labors shall be painful.

The process of parturition is purely a physiological one, as much so as the process of urination and defecation. The latter especially is closely allied to parturition in its mechanism. I do not believe the one should be more painful than the other.

Notwithstanding it has been asserted by travellers and residents among aboriginal tribes, that the women of such tribes *do* suffer from the "pangs of childbirth," I do not believe that their so-called sufferings compare in nature and intensity with those of women in a state of civilization.

The fact is, that there are very few tribes on the globe which can be said to be now living in a *natural* state. The slightest contact with civilization modifies their habits and customs, and every slight modification tends to alter their physiological functions.

I have received positive and trustworthy information from those who have spent many years among our native tribes, that they know of hundreds of instances where the process of childbirth was not attended by other than *painless* expulsive efforts. And more: nearly every physician of large obstetric practice has attended cases in which the woman was confined in so short a time, and with so little suffering, that the labor could be termed *painless*.

In my own practice I have known many instances where *pain*, as properly understood, was not present. The labor was attended by bearing-down, expulsive efforts, and some vascular and mental excitement, nothing more. Therefore *normal* labor may be termed a *painless function*.

So long as the human race can never return to a state of existence such as we picture as the ideal, so long will a large proportion of women be subject to dystocia.

What are the elements to be taken into account in the *preventive* treatment of functional dystocia?

If possible, we should first ascertain if any *structural* obstacle exists. If we find none, we should then inquire into the history of the patient.

If a *primipara*, we should get her menstrual history, as this may be some guide as to the probable nature of the labor.

"Some writers," says Cazeaux, "have attempted to establish a relation between the phenomena that precede or accompany the menstrual discharge in the non-gravid state, and the activity or slowness of the contractions of the womb during the labor: for they say, should the periodical flow be difficult, laborious, and painful, and the patient be tormented every month with violent colicky pains, either before or during her terms, the irritability of the uterus, and the energy of the contractions will almost invariably be excessive in the hour of childbirth; but, on the contrary, there is reason to anticipate the occurrence of slow and feeble pains when the woman is advised of the return of her menses only by the appearance of blood, and when they pass off without suffering."

There is much truth in this theory, for it has been observed that those women who were the subjects of violent dysmenorrhœa, usually have very painful labors, especially the first; and that the subjects of menorrhagia were more liable to postpartum hæmorrhage. In the case of *multipara* we may judge of the probable nature of the dystocia by the history of her previous labors. In this way I have known of hundreds of cases where the sufferings of the woman have been greatly palliated, and even almost prevented.

Not only should we inquire into the history of the previous labors, in their relation to the uterus, etc., but into the condition of the other organs at the time of labor.

If in previous labors or pregnancies, the heart, lungs, kidneys, or bowels were in an abnormal condition, we should select our remedies to accord with the abnormal condition of those organs.

In this connection it should be stated that the commonly received notion among women, that the distressing and painful sensations occurring during pregnancy must be borne without applying for relief, *because they occur during pregnancy*, should be abandoned, and it should be the duty of the physician to disabuse the minds of his patients of this erroneous belief.

Normal pregnancy should not be attended by painful symptoms; consequently all such are abnormal, and demand relief, and the carefully selected homœopathic remedies will generally give relief.

The principal medicines indicated for the disorders of pregnancy, or as prophylactics of dystocia, are:

Arnica, Æsculus, Aletris, Bromides, Caulophyllum, Calc. c., Cimicifuga, Collinsonia, Digitalis, Ferrum, Eupatorium purpureum, Gelseminum, Gossypium, Helonias, Ignatia, Nux vomica, Pulsatilla, Secale, Scutellaria, Trillium, Senecio, Sepia, Ustilago, Viburnum, and *Veratrum viride.*

In order to have a clear understanding of the nature of these remedies and their indications, we will consider the

GENERAL DISORDERS OF PREGNANCY. 287

various causes and conditions which relate to dystocia and mention the remedies indicated for such conditions.

Arnica causes a depressed state of the nervous system, especially the cerebro-spinal and the vaso-motor, leading to a stasis of the capillaries. When pregnant women complain of soreness of the muscles, lameness, and weakness of the limbs, and such weakness of the capillaries that discolored spots appear upon the skin after slight pressure, the external use of Arnica-water, and the simultaneous use of Arnica internally, is of great value in preventing painful and difficult labors. It also prevents that extreme soreness and tenderness of the uterus and genital passages which occurs even after normal labors.

Æsculus will remove many of the symptoms of painful pregnancy, particularly the lameness in the sacro-ischiatic ligaments, but it is especially useful for the disorders of the rectum, which often cause great discomfort and lead to dystocia. It removes the tendency to hæmorrhoids if given in time, and tends to prevent the occurrence of painful and protruding piles after labor. Premature labor is often induced by rectal irritation, and in such cases Æsculus will prevent that accident.

Aletris is a powerful tonic to the muscular system, and the organs concerned in the process of nutrition. It is useful if the woman is weak, cannot endure exercise, has loss of appetite, and a constant sensation of weight and pressure downward in the pelvis. A few drops of a low dilution before meals will often restore the appetite and nutrition in a brief time. If the liver is torpid in addition to the above symptoms, and the woman has gastric catarrh, Hydrastis will be useful.

The *Bromides* are very useful agents in the treatment of the pregnant state.

Plethoric patients are often troubled with an undefinable nervous erethism, accompanied by hallucinations, morbid impulses, even sexual appetite in excess, horrible dreams, or exceeding wakefulness. The ordinary household cares worry them exceedingly, they are restless and sleepless, have rush of blood to the head, vertigo, fainting, etc. In such cases I have seen such excellent soothing results from the bromides, with such an evident improvement in the general mental and bodily health, that I sincerely advise their use.

For the indications, consult the provings in "*Symptomatology of New Remedies.*"

For general use I prefer Bromide of *soda*, but if there is decided intracerebral congestion, with pain in the occiput, and hemicrania, the Bromide of *ammonium* is preferable, while for purely mental aberration the *potash* salt is the best. If the tendency to congestion of the brain is attended by scanty urine, the Bromide of *lithia* acts promptly. When hysteria, faintings, and tendency to spasms are imminent, the Monobromated *camphor* is of great utility. The dose of these preparations must be varied with the exigencies of the case. I have never seen good effects from them in the high attenuations.

From *one* to *ten* grains, repeated at suitable intervals (one to six hours), will be found to remove the above-named morbid manifestations.

In cases of threatened cerebral congestion, with cold hands and feet, flushed face, distended carotids, vertigo, incoherent speech, etc., I have often given twenty grains of the Bromide of *lithia*, with the result, as I believe, of saving life or preventing severe illness.

In the most violent cases of hysteria, and hysterical spasms, from one to five grains of the 1^x trit. of Bromide of camphor often arrests them, if repeated every ten or fifteen minutes. During the latter months of pregnancy some women are so oppressed by cerebral congestion that puerperal eclampsia is imminent. I believe this fearful accident may be warded off by the bold and judicious use of the bromides. Even if uræmia is present, the bromides are not contraindicated, for they aid in the elimination of the poison, acting as diuretics, especially when their use is associated with the administration of Apocynum, Apis, Cantharis, Helonias, etc.

Caulophyllum is one of those indigenous remedies which, from the earliest occupation of America by the whites, has been known to be used by the aborigines as a preventive of difficult labors. Nearly every tribe use this, or some analogous plant, the woman drinking a weak decoction of it a few weeks prior to expected labor. We have not yet anything approximating a complete knowledge of its action on the human body; the provings are meagre and unsatisfactory. But from the large mass of clinical evidence, we deduce that it acts upon the smaller muscles, the uterus, and perhaps all other hollow organs. It appears to act on the muscular fibres which engage in expulsive action. Its primary action is to exalt the irritability of such muscular fibre, its secondary to depress. We may prescribe it in the absence of special indications, in primipara, when the menstrual periods have been very painful, from *spasmodic* action, and where the pains of the dysmenorrhœa have been like labor-pains, and the discharge small in quantity and short in duration. Or in multipara when the previous labors appear to have been violent, painful, the pains spasmodic, regular, but with rigidity of the os and soft parts. In very many cases I have prescribed the Caulophyllum for these general indications, and with the happiest results, the following confinement being natural and comparatively free from pain.

During the last fifteen years I have published in the various journals of our school many cases corroborative of its power over dystocia. Many such cases have been reported by physicians of our school in all parts of the world.

Besides its value in preventing excessively painful, tedious, and slow labors, it has other powers equally useful. It mitigates, and often altogether prevents those annoying and distressing "false pains," so common to the women of our cities and large towns. These pains are generally located in the hypogastrium, they appear at nearly regular intervals, the paroxysms coming on every night for weeks, and the pains recurring every few minutes or every hour. They often extend into the limbs, up the sides, and into the abdomen. Some women are so irritated by these pains that premature labor is induced by them. In these cases not only does the Caulophyllum arrest the pain, and make the last few weeks comfortable, but it gives an easy labor, and according to the testimony of some of my colleagues, protracts by several days or a week the duration of pregnancy.

The *dose* of Caulophyllum is an important consideration. It is claimed that good effects have been obtained from the third dilution, or the sixth trit. of Caulophyllin. In a few cases I can verify these claims, but in a majority of instances I prefer the lower attenuations. The violently spasmodic, very painful effects of the drug are primary effects—the atonic, paralytic effects, secondary. Therefore, while the middle attenuations will act curatively in very painful, spasmodic conditions, due to great irritation, the lowest preparations, even the crude drug, may have to be used for atonic conditions. Physicians of the Eclectic school habitually prescribe the tincture in five-drop doses, or the active principle in doses of fractions of a grain, for weeks before confinement, with no alleged bad results. I believe I have seen dystocia, and painful symptoms during pregnancy, warded off by the third attenuation; but I have also seen the happiest effects from the 1^x, 2^x, and even drop doses of the tincture, repeated every two, three, or four hours. It is well to begin the use of Caulophyllum a month or six weeks before full term; or six or ten weeks, if the woman fears a premature labor at the seventh or eighth month—a dose to be taken once or twice a day. It has been known to enable women to carry children to full. term, who habitually had premature labors.

It may be well to mention a few practical hints as to the methods of administration. Pellets may be employed if the dilutions are used; large pellets will each absorb nearly a drop of the alcoholic mother tincture. The tincture or first dilution in drop doses often causes an unpleasant burning in the throat. This may be avoided by prescribing it in simple syrup or gum-water. The low triturations of Caulophyllin have the same effect, which may be avoided in the same manner. An elegant method of administering a definite dose of the active principle is in the form of a sugar or gum-coated pellet or granule, manufactured by several firms in this country. These granules contain quantities from the 1-10 grain up to 1 grain. The former is best for use during pregnancy; the latter to facilitate labor.

Cimicifuga is a near analogue of Caulophyllum, but has some notable differences in its method of action. 1. It acts upon the cerebro-spinal system, giving rise to spasmodic, neuralgic, and myalgic pains. These pains and spasmodic motions are not symmetrical, but irregular, and tend to run into choreic movements. 2. It affects specifically the large muscles, and has the power of causing a variety of muscular pains, such as rheumatic, myalgic, and spasmodic. 3. It affects the brain and mind, causing congestive headache, melancholy, and mania.

All these conditions above named are well known to be concomitants of the pregnant state, and lead to the most troublesome forms of dystocia.

Dr. Meigs (*Obstetrics*, page 260), asserts that a large proportion of the cases of "false pains" occurring during pregnancy are due to uterine rheumatism. He gives a clear and graphic description of such cases, and his opinion is very valuable. I have observed that many pregnant women who are troubled with "false pains," and who have very painful labors and severe after-pains, have been subject to dysmenorrhœa, which was undoubtedly of a rheumatic character.

Now, of all remedies for rheumatic affections of the uterus, none are of greater value than Cimicifuga. Its only rival is Guaiacum, Salicin, or possibly Viscum album.

In many instances of abnormal pregnancy it is superior to Caulophyllum, because of its wider range of action, and greatly superior power. If the student will consult the latest full pathogenesis of Cimicifuga, he will see the special indications fully set forth. My records of cases contain many wherein this remedy has removed profound melancholy, insanity, and a condition simulating delirium tremens. These women might have been immured in lunatic asylums but for its curative power.

A frequent and distressing attendant of some pregnancies is the well known "pain in the left side," under the left breast. This pain may shift to the left ovarian region, or upward to the left arm, and even change to the vertex.

All these pains promptly give way to a lower attenuation of Cimicifuga. Many physicians, especially those of the Eclectic school, value Cimicifuga more highly than Caulophyllum as an agent for the prevention of dystocia. The Eclectics use it on general principles; they claim that it brings about a normal relaxation of the muscular structures, which if "rigid," would obstruct the progress of natural labor; that it gives tone to those muscles which have an expulsive function; and moreover, that it gives the nervous system that strength and tone so requisite in such cases. When speaking of the immediate treatment of dystocia the indications for its use will be more fully set forth.

As a general prophylactic of dystocia, it should be prescribed to be taken several weeks before full term, or otherwise, as set forth under Caulophyllum; the active principle, *Macrotin*, can be given in the 2^x or 3^x triturations.

Collinsonia has many points of resemblance to Æsculus. It acts upon the same organs and tissues, but has in addition a *toning* power, especially over muscular fibre. It will remove the obstinate constipation and hæmorrhoids of the last months of pregnancy. It does this by restoring the normal irritability of the muscular coats of the intestinal canal. It is well known that during pregnancy there occurs a *normal* or physiological hypertrophy of the heart, to compensate for the increased demand for a more powerful circulation. If the pregnant woman has a neurosthenic diathesis, this hypertrophy will be attended with an increase of the action of the heart far beyond its proper bounds. The pulse will reach 120, but regular, when it should be 80 or 90. If the conditions above named all occur together, during the last month of gestation, dystocia would be sure to result, followed by an unfavorable puerperal state. Collinsonia, in the attenuations from the first to sixth will be found efficacious if it is used patiently and persistently.

Digitalis is a far more potent medicine than the last named, and in many pregnancies is absolutely indispensable. By its judicious use a woman may be carried to full term, when suffering from cardiac disease, when without it she might not survive to reach the last months, or, if surviving, suffer miserably.

If a woman enters upon pregnancy with a "weakened heart," whether from functional or organic disease, the physiological "strain," which it is called upon to undergo, soon merges into abnormal conditions. In the so-called "irritable heart," when its action is *irregular, intermitting, weak*, and *excitable*, no remedy compares with Digitalis in its controlling power. The patient, who before its use could not go up stairs, or take needed exercise, without distressing palpitations and dyspnœa, soon finds herself more comfortable in every respect. Digitalis acts upon the heart as does Cimicifuga upon the uterus, regulating irregular action and imparting strength through its nervous supply.

A woman who goes through gestation with any organic heart disease (except hypertrophy with thickening) is liable any moment to exhaustion, and even death from cardiac failure. But if the heart is kept *supported* by Digitalis, no such accident will occur under ordinary circumstances.

A weakened heart implies irregular and generally deficient blood-supply in every organ of the body. In the brain we have anæmia or venous stasis, so also in the lungs, uterus, and kidneys. The kidneys usually suffer most, for the irregular blood-pressure in those organs lays the foundation for those conditions known as Bright's diseases, and consequent uræmia. Even if the heart is not complained of, and the patient has irregularity in the quantity of urine, at one time scanty and dark, at another profuse and watery, and especially, if œdema of the face or feet is present, Digitalis should be prescribed. But it is useless to give Digitalis unless the proper *dose* is selected. The exclusive high attenuationist will not see any good effects from it in the above conditions, for they *all belong to the secondary effects of the medicine*, and the scientific law of *dose* calls for material quantities. As a rule, a grain or two of the 1^x Digitalis *leaves*, or the 2^x of Digitalin, or two or three drops of a good tincture, repeated three or four times a day, will bring about good results. In rare cases the 2^x will suffice. These doses can be continued days and weeks with none other than favorable effects; no "cumulative action" is ever seen from these doses, and in these conditions. If *anæmia* is present, and the patient fails in obtaining good blood from food, the use of Ferrum should always be associated with Digitalis. These two medicines act very harmoniously in such conditions, the iron cures the malnutrition, and soon enriches the blood, and the Digitalis insures its normal distribution by means of a strengthened heart. I prefer the 1^x trit. of *Ferrum met.* or *Ferrum lact.* to any other preparation. Give the Digitalis before meals and the Ferrum after. Digitalis will also prevent hæmorrhage after labor, when there is a predisposition to it. It acts on the uterus as it acts on the heart, inducing firm contractions.

Lycopus virginicus is a near analouge of Digitalis. It will control a weakened, irregularly acting heart, with dyspnœa and cough from pulmonary stasis. It has not, however, the dropsical symptoms of Digitalis, nor the scanty urine. On the contrary, the urine is profuse—a condition simulating diabetes.

Helonias has long had a reputation as a true "uterine tonic." The few provings made, seem to show that it resembles Cimicifuga Aletris, Ferrum and Pulsatilla. Certain it is, that all the generative functions when disordered, seem to improve under its influence. Not only does it exert a profound influence over the uterus and ovaries, but the *kidneys* are included in its sphere of action. It has cured many cases of albuminaria and desquamative nephritis, and conditions simulating Bright's diseases, and I predict it will be found one of our most efficacious remedies against urœmia. If a pregnant woman in the last months becomes enfeebled, pale, anæmic, dyspeptic, with albuminous urine, excessive or deficient in quantity, and has great depression of spirits, the Helonias should be precribed in the 2^x or 6^x dilutions (or triturations of Helonin), several doses a day until her condition has decidedly improved, and then repeated less often, up to the day of her delivery.

Gossypium is a medicine about which there is still an atmosphere of doubt and incredulity. Many physicians believe it to be a valuable uterine tonic, others deny its power altogether. The provings found in "*New Remedies*" seem to show that it has some action on the uterus. It is probable that the fault lies in the difficulty of perserving intact the virtues of the root in any discovered preparation. The fresh root is certianly a uterine motor, and has been known to bring on labor, or facilitate difficult parturition. It has been used for the same purpose as Caulophyllum, as a preparatory agent for the purpose of bringing about natural and easy labor, and with alleged success. The dose is usually from five to fifteen drops of the tincture daily.

Pulsatilla.—I have for many years been convinced that the uterine affinities of this medicine have been greatly overrated. It may cause uterine blennorrhœa, venous congestion, and neuralgia, but I do not believe it to be a uterine motor. It removes many of the sufferings of pregnancy, such as toothache, prosopalgia, vomiting, indigestions, intestinal derangements, and anomalous nervous affections of the generative organs. In this way, if selected according to its symptoms, it may prove a remedy to ward off dystocia. In its pretended power to change abnormal presentations of the fœtus in utero, I have not the slightest confidence. We know that spontaneous version is a common occurrence, and that it happens apparently when most demanded, and at the very time when Pulsatilla would most likely be given for that purpose. Of its use during painful and difficult labor I shall speak in another place.

Secale ought, by its undisputed power over the impregnated uterus, especially during the last months of gestation, to prove an efficient remedy against dystocia, when prescribed previous to that event. I am not aware, however, that it has been used as such, except in a few instances, and nearly all of those in my own practice. These cases were all characterized by the peculiar cachexia belonging to Ergot, in which there is a strong tendency to failure of the vital powers, a predisposition to hæmorrhage, and a lack of vitality in the fluids of the body. If a pregnant woman becomes feeble, emaciated, with dry shrivelled skin, cold hands and feet, sensation of great weight in the uterus, and worrying false pains, and feels as if she should be sick any day during the last month, Ergot in small doses is the remedy, and especially so if her previous labors have been slow, feeble, and followed by hæmorrhage and long-lasting, fetid lochia.

This powerful uterine motor ought to prove one of the best of all remedies against dystocia. It is necessary, however, that we should have a correct understanding of its method of action, in order to prescribe it successfully.

Primarily in large doses it causes violent, persistent, tonic contractions of the uterus, with expulsive efforts, also vaso-motor spasm of its bloodvessels, but this is followed by a corresponding depression of the motor power of the uterus, with vaso-motor paralysis. Thus, *secondarily* we have uterine atony with tendency to passive hæmorrhage. This gives us a clue to its true curative action as a preventive of dystocia. If we have reason to fear that the woman will be confined prematurely from want of tone of the uterine muscular fibre, or, if she goes to full term, only to have a slow, distressing labor, with deficient pains, lack of expulsive power, tendency to hæmorrhage from deficient contraction, and finally sub-involution of the uterus, Secale, in a low dilution, may be given every day during the last month, with every prospect of success.

(The dose I advise in such cases is one to five drops of Squibb's Fluid Extract three times a day, up to the day of confinement, beginning ten or fifteen days previously.)

Be *sure*, however, that the preparations used are reliable, for of all medicines Ergot is most likely to be inert. I prefer Squibb's preparations, or the French Ergotin, and am sorry to say that the tincture ordinarily sold cannot be trusted.

Ustilago.—This fungus, although a near analogue of Ergot, cannot be said to be identical in action. It has more control over the unimpregnated uterus, and over the impregnated in the *early* months. It is of more value as an anti-abortive than as a preventive of premature labor. It is especially indicated in a certain atonic condition of the uterus, when there is a passive hæmorrhagic condition, the blood being black, grumous, or muddy. There is but little pain attending this symptom, only a sense of weight in the uterus and much general prostration. During the last months of pregnancy, these symptoms are sometimes annoying and give rise to much anxiety. This anxiety is not without cause, for they portend a slow, difficult, and dangerous labor, on account of the atony of the uterus. During labor, hæmorrhage difficult to control, may occur, owing to deficient contractility. This condition, however, may be prevented by giving the lower triturations, several doses a day, for a week or more previous to confinement.

Viburnum.—This medicine, which has of late become very popular in the treatment of dysmenorrhœa and allied affections of the uterus, has been called by the expressive name of cramp-bark, from time immemorial. I was, I believe the first in our school to call attention to the peculiar properties of this medicine, although it has been used in domestic practice for a century. Its common name perfectly expresses its remedial power. Its action on the nervous system gives it a decided curative power over *painful contractions* of muscles, particularly the flexors of the limbs, and those engaged in expulsive actions in hollow organs. Like Caulophyllum, Cimicifuga and Helonias, it had a reputation among the aborigines and early settlers of this country, as a *preventive of painful labor*. An infusion of indefinite strength was drank daily for a few weeks previous to expected confinement. Empirical experience has enabled us to define its range of action, and its curative power. It is indicated when the patient has been the subject of exceedingly painful dysmenorrhœa, or uterine cramps, accompanied generally by cramps in the limbs, abdominal muscles, or other parts of the body. In such cases it is to be considered probable that the woman will have a very painful labor due to cramp-like contractions of the uterus, cramps in the limbs, etc. Associated with these symptoms will appear excessive erethism of the nervous system. All species of the Viburnum, which I have examined, contain a large percentage of Valerianic acid, which gives it great power over abnormal conditions of the nervous system. It is especially indicated for hysterical women, and when we have reason to fear that the labor will be complicated with hysterical spasms. Some cases of puerperal eclampsia are purely hysterical, and may be prevented by the previous administration of the Viburnum opulus, or Viburnum prunifolium. The latter possesses great power in preventing miscarriage and premature labor, and its use should never be neglected when painful cramp-like pains attend any period of pregnancy especially the last months.

The dose found most efficacious varies from a teaspoonful of the tincture, or fluid extract, to single drop doses of the mother tincture, or the 1^x dilution. These doses are perfectly harmless, for it is a uterine *sedative*, not a uterine motor-excitant. It is the opposite of Caulophyllum, Ergot, or Ustilago. These doses should be repeated several times a day for weeks before labor, or every hour or two during the pains.

Veratrum viride.—My own experience and observation has convinced me that this medicine is invaluable in some cases of threatened and apparently inevitable dystocia.

In some constitutionally plethoric women the plethora increases to a great extent during gestation, and this condition is increased to a dangerous extent by the physiological cardiac hypertrophy.

An enormous quantity of blood is manufactured, and the body is not only fed to excess, causing great increase in its weight, but all the important organs are engorged. In such cases it may be expected that *convulsions* will occur upon the slightest exciting cause.

Apoplexy, both of the brain and lungs, is a common occurrence in such an abnormal state. It is well known that cutting off the supply of blood-making food will not always reduce the plethora, and venesection being virtually abandoned by the old school, is never seriously thought of by ours.

We have, however, in Veratrum viride a remedy which, associated with a low diet, will, in nearly every case, arrest the inordinate plethora. This it does by slowing the action of the heart, and lessening its abnormal force. Under its influence I have known the bodily weight decrease several pounds a month, with general relief to the engorged head and lungs. It also decreases the abnormal arterial pressure in the kidneys, and is of great value in preventing serious renal congestion. Its sedative action on the cerebro-spinal system and the convulsive centre in the brain, tends to prevent the occurrence of puerperal eclampsia, for which accident it is almost specific.

The proper dose varies according to the susceptibilities of the patient, ranging from one drop of the first dilution (1^x) to five drops of the mother tincture, repeated three or four times daily.

It should be given largely diluted with water, and during its administration, strong tea and coffee, or spirits, and animal food should be prohibited. The use of the alkaline mineral waters, Vichy, Seltzer, or Kissingen, aids its action.

Hydrotherapy.—The careful and judicious application of some of the measures used in the so-called water-cure system, are very efficacious in removing many of the ailments of pregnancy and thereby preventing dystocia.

Very many instances have come under my observation wherein women who have failed to carry the child to full term, owing to laxity of the uterine tissues, or irritability of that organ, have been enabled to give birth to a child at full term by means of the sitz-bath, etc.

Other cases I have known, where women have had tedious, difficult and dangerous labors, when a residence for a few months in a Water-Cure establishment would seem to remove all the abnormal conditions, and their next labor would be easy and natural, nor followed by unpleasant sequelæ.

The *home*-treatment of pregnant women is not carried on as successfully as at a "Cure," unless the patient has every means at her command, and also possesses a persevering and patient disposition. It is probable that the *rest* and freedom from domestic care, the systematic exercise and the excellent healthful diet, have much to do with the cure of such patients. But when a residence in a "Cure" is not convenient, or feasible, the physician can give proper directions for the home treatment.

The three most useful hydrotherapic appliances are: the *sponge friction-bath, sitz-bath,* and *wet bandage.*

In the majority of cases all three will have to be combined, especially when there is general and local debility.

The sponge-bath, associated with smart rubbing or friction, with the hand or brush, is a powerful tonic to the system, providing the patient is *strong enough to bear it.* In cases of feeble and delicate women it should be used with great caution, not repeated too often, or its use prolonged beyond a few minutes. It is not necessary to expose and bathe the whole body at once, but a portion at a time, leaving the rest clothed.

Alcohol or spirits should never be put in the water, unless they leave a warm glow, and a soft, supple skin.

In very much emaciated subjects, the use of oil instead of water is a powerful means of *nutrition*, and should not be neglected. It should be thoroughly rubbed into every portion of the body, and repeated every day or two, with a general water-bath once a week. Even dry rubbing by a healthy person is a tonic to the muscular and nervous system if it is agreeable to the patient.

The *sitz-bath* (hip-bath) is best adapted to relieve local congestions and weaknesses of the sexual organs. Of all the water-cure appliances this is the most popular among weak and debilitated women. Properly used, it acts as a tonic to the circulation of the pelvic visceri.

The *modus operandi* of the sitz-bath has been variously explained. My views of its action may be briefly stated. The *very hot* sitz-bath has nearly the same effect as the *very cold*. Both primarily contract the external bloodvessels and determine an unusual flow of blood to the internal organs of the pelvis. But these extremes are rarely useful. The temperature of the water should rarely exceed 100°, or fall below 60°. At this temperature they rather equalize the pelvic circulation, and have a *toning* effect. They have doubtless a stimulating effect on other nerves than the vaso-motor, as it is well-known that hot or cold sitz-baths increase the power of uterine contractions.

But leaving out all theoretical explanations of their value, we know that the regular and persistent use of sitz-baths do greatly assist in the cure of chronic uterine disorders, and avert threatened miscarriage and prevent habitual abortions. We also know that they are a great comfort to delicate women in the last months of pregnancy, relieving false pains and many other abnormal sensations. I have known many feeble women use the sitz-baths daily all through pregnancy, up to the very day of confinement, and more, I have known them pass the hours of labor in a sitz-bath. These women had previously suffered intensely all through their pregnancies, but they all declared that these baths prevented similar sufferings and even mitigated greatly the pain of labor. The *temperature* of the water should be made to suit the feelings of each patient. Let each be her own judge.

The *duration* of the bath may be ten or thirty minutes, and this too may be generally left to the sensations of each patient.

The *depth* of the water should be just sufficient to cover the hips and rise to the crest of the ilia.

The *wet bandage* is another method of applying water, pure or medicated, that should not be neglected. There are patients who are not able to use hip-baths, or are not pleasantly affected by them, yet these patients suffer from some or all of the following symptoms, namely: sensations of soreness, tenderness or sensitiveness of the lower abdomen, weight, dragging and bearing down in the pelvis, constant tearing or burning pains in the sacral region, heat and fulness in the uterine region, and many other symptoms pointing to passive stasis of uterine circulation and weakness of its muscular structure.

In such cases the regular use of the wet bandage every night, and even day and night, is followed by the best results, namely: the gradual removal of all the above symptoms. The bandage should be of light towelling, double or treble, and covered by one thickness of flannel or cotton drilling, and should be a foot wide. It should be wrung out in water cool or warm to suit the feelings of the patient (*cool* is advised), and changed as often as it becomes too warm or dry. The water may be medicated with Aconite, Arnica, Nux, Hamamelis, or Cimicifuga, in proportion of one-half ounce to a gallon.

CHAPTER II.
SPECIAL DISORDERS AND ACCIDENTS OF PREGNANCY.

NAUSEA AND VOMITING.—In many cases of pregnancy, intense nausea, or severe and painful vomiting renders the life of the woman almost insupportable. Not only does it cause great physical discomfort, but it interferes greatly with the general health and enfeebles the system, rendering the woman poorly prepared for the exigencies of labor. Until within a few years it was supposed that little or nothing could be done by the physician to alleviate this symptom. But with the advance of our knowledge of the pathological conditions which may be incident to the pregnant state, and the discovery of the law of *similia* and the pathogenetic qualities of medicines, much has been done to relieve the above symptoms.

But, as in nearly all pathological conditions, we must ascertain the *cause* of the vomiting or nausea before we can remove it. Until lately the theories which were put forth to account for the presence of this vomiting, were vague and unsatisfactory. The general belief, however, was that it was caused by "reflex irritation," and that the source of irritation was to be found in the uterus; but as to the exact *condition* of the uterus which caused this reflex irritation no satisfactory explanation was given. The assertion that the vomiting was caused by the presence of the impregnated ovum, or growing embryo, or enlargement of the uterus, are all untenable, because only a small proportion of pregnant women have sufficient nausea to be unpleasant, and many none at all. If either of the above conditions caused the vomiting, *all* pregnant women would suffer from it. After carefully analyzing all the most recent theories of the causes of vomiting during pregnancy, I believe there are three conditions only to which it is due, namely:

(1.) The presence of the impregnated ovum.
(2.) Flexions and versions of the uterus.

(3.) Constrictions of the cervix uteri.

(1.) Many women assert positively, and many gynæcological authorities place reliance on their testimony, that nausea sometimes occurs immediately after a successful intercourse, *i. e.*, an intercourse during which the semen effects an entrance into the uterus.* We also know that the mere passing of the sound into the uterine cavity, or the introduction of a few drops of fluid, will in some women, cause nausea and vomiting. The ordinary nausea of pregnant women, however, usually occurs about the fifth or sixth week, at a time when the embryo begins to exert by its growth some pressure upon the parietes of the uterus; but this ordinary nausea usually passes away in a few weeks.

I ought to add that it is Dr. Hewitt's opinion† that not only the severer forms of vomiting of pregnancy, but even this supposed normal nausea of the early months, is due to slight flexions of the uterus. He thus explains it: "It is the fact that the patient generally experiences the symptoms in question on first rising in the morning, or while dressing. Why is this? Is it not because the body of the uterus falls a little downward in obedience to the law of gravity, thereby producing a slight flexion and a compression of uterine tissues at the seat of the flexion. During the first three and a half months such a temporary flexion is possible, because the uterus is still in the pelvis. Generally after that time it rises out of the pelvis, and flexion to more than a very slight extent is no longer possible (but *version* is of frequent occurrence). Is it not the fact that for the most part the liability to nausea and vomiting ceases at precisely this period. It is also a fact, which will be confirmed by all who make the experiment, that in ordinary slight cases of nausea and vomiting, by ordering the patient to remain absolutely in the horizontal position the symptom ceases." Dr. Hewitt, perhaps, carries his theory too far, for it is well known that by taking her breakfast in bed, or drinking a glass of champagne before rising, the nausea is ameliorated. Still it

---

* See Pamphlet by Dr. M. A. Pallen, heretofore quoted.
† Diseases of Women, p. 436.

is a suggestive fact, that the remedies most useful in flexions of the uterus, namely: Sepia, Lilium, Nux, Puls., etc., are often our best remedies in this nausea, although Ipecac. is generally very useful, for other reasons.

(2.) That the really severe and painful vomiting which renders some pregnancies so distressing, is caused by *flexions and versions of the uterus*, there is now but little doubt in the minds of well-read physicians. We are chiefly indebted to Dr. Grailey Hewitt, of England, for the discovery of this fact. In a paper presented to the Obstetrical Society of London, 1871, he propounded his views on this subject, which are further set forth as follows in his *Diseases of Women*, page 429. His argument is (*a*) that nausea and vomiting are associated with pregnancy. (*b*) Nausea and vomiting are associated with disease of the uterus; (*c*) but nausea and vomiting are not *always* present in cases of pregnancy, nor are these symptoms always present in cases of uterine diseases. (*d*) It is probable that the nausea and vomiting of pregnancy and that from uterine disease, are both due to a similar cause. (*e*) Nausea and vomiting are very common in *flexions of the uterus*, especially when the flexion prevents the escape of any fluid in the cavity of the uterus.

Dr. Hewitt arrives at the conclusion that the "*sickness of pregnancy* is due to the combined effects of the increasing distension of the uterus and an associated flexion of that organ." He offers a large array of proof in support of this theory, and I must say that the testimony which he brings forward seems indisputable, especially so in view of the fact, that for many years my experience substantiates it. Dr. Hewitt says: "Having occasion to treat cases of sickness in young unmarried women suffering from flexion, it has been observed by me that when those patients marry and become pregnant, the sickness observed is liable to be unusually severe and troublesome." He also observes that "cases which have come to me in the course of consultation practice, and when the symptoms of sickness have been still more troublesome, I have always recognized an abnormal condition of the uterus as regards its shape."

Treatment.—In excessive nausea and vomiting from actual flexions of the uterus, the treatment should be more *mechanical* than medicinal. I do not think it possible that any medicinal agent can restore the bent uterus to its normal shape, without the aid of *posture* and *mechanical supports*. Says Hewitt: "Over and over again have I observed these troublesome symptoms disappear at once on applying mechanical treatment for the restoration of the uterus to its proper shape, whereby the congestion and the irritating pressure on the uterine tissues are removed. Mere attention to the position of the body often suffices to relieve the patient. The horizontal position on the back in cases of *anteflexion*, the prone position in cases of *retroflexion*, unaided, give great relief, a relief which is more effectual if conjoined by suitable internal appliances for aiding the restoration of the uterus to the proper shape."

It may be interesting in this place to quote a few cases confirmatory of this theory, reported by physicians of our own school. The following are reported by R. N. Foster, M.D., Professor of Obstetrics in the Chicago Homœopathic College:

"Case I.—Mrs. ——, æt. forty, had advanced to the fourteenth week of her second pregnancy. Had been pregnant for the first time just fifteen years ago, and had then miscarried at the twelfth week—the miscarriage being due to persistent and intractable vomiting. The history of this, her second pregnancy, had thus far been an exact repetition of the first. For ten weeks she had been confined to her bed by frequent vomiting, which was made worse by sitting up or by standing or walking. Her appetite was good, but a few minutes after eating any kind of food she vomited it.

"She had been under 'regular' treatment, but without any mitigation of the annoyance, besides she was growing weak and emaciated, and dreaded a second miscarriage.

"An examination detected the uterus lying, thoroughly retroverted, in the hollow of the sacrum, which the patient informed me had also been the case in her previous pregnancy. The retroversion was probably a consequence of the increased weight of the womb, as the ready mobility of that organ would indicate that the position had been recently assumed. Moreover, the pelvis was well formed and roomy, and the replacement of the womb was therefore free from difficulty. It was accomplished by alternately lifting the body and fundus up out of the hollow of the sacrum, and pushing the cervix back toward the coccyx. This was done gently, and the organ thus gradually restored to its proper position.

"To retain it there an inflated ring-pessary was introduced and placed precisely as in cases of prolapse, *i. e.*, so that the os rested in the central opening of the pessary.

"The next day the patient resumed her household duties, the vomiting having ceased at once, and she is now approaching the fortieth week of pregnancy, and is in perfect health.

"In retroversion of the womb, when the condition is not complicated by adhesions, the best 'repositor' is the index finger, or, if this does not suffice, another inch of progress may be gained by the introduction, within the vagina, of the middle finger also. The elevation of the fundus ought immediately to be followed by pressure on the anterior surface of the cervix, directing that portion of the organ back toward the coccyx. This latter manœuvre will often succeed when the former alone fails."

"Case II.—Mrs. ——, primipara, had been married three months; had 'gone over her time' two weeks, a circumstance by no means unusual with her, and therefore creating no suspicion of pregnancy. But just at this time began an era of terrible vomiting. If she ate anything she vomited; if she drank anything she vomited; if she neither ate nor drank she vomited. She retained her food long enough to obtain some little nourishment from it, nevertheless in three weeks she was reduced to a condition of positive danger from starvation. She was pallid, weak, emaciated, with a pulse of 100, daily growing more rapid, and occasional streaks of blood in the substances vomited. Moreover, she was a delicate woman of consumptive proclivities.

"She had meanwhile been constantly under treatment, dietetic and medicinal. She had numerous friends, and every one of them had a new diet, which was tried, and each discomfited friend retired to make room for the next. Her treatment was somewhat mixed, ranging all the way from Ipecac. to the Oxalate of cerium, in 'regular' doses. The reader may take it for granted that I simply gave everything 'according to the symptoms,' and *not* according also, and that my patient daily grew worse.

"At length I asked for an examination. Everything in the pelvis seemed normal, both as to condition and location. The womb was no lower in the pelvis than is usual in perhaps one third of our cases examined, whether pregnant or non-pregnant. Nevertheless I determined to introduce an inflated ring-pessary, by which means I would at least alter the position of the womb, and bring what little pressure there was to bear on new points and in new directions. Possibly an oversensitive nerve might thereby be relieved, and this reflex irritation stopped.

"Therefore the pessary was introduced, and the vomiting ceased immediately; occasional slight nausea was experienced during the next two months, but not so as to cause the patient any real discomfort. The pessary was not felt by the patient at all, and was retained for three months, when it was removed. She is now in her thirty-sixth week and well."

"CASE III.—Dr. A. W. Woodward lately reported to me a case equally intractable with the two preceding, when I related to him this 'story of the two pessaries,' thereby inducing him to try the same experiment. The result in his case was a total failure, wherefrom we may reasonably conclude that inflated ring-pessary is not a "specific" in this disease, and indeed we do not as such recommend it. But the second case suggests as one cause of the vomiting in these cases a hyperæsthesia of the uterine nerves, of such a character and so located that simply altering the position of the uterus may chance to settle the whole difficulty. The pessary is therefore worthy of trial in such cases, especially as the experiment, if not successful, is at least perfectly harmless. Besides, how can we by medication cure a case which is of such a nature as to yield at once to this simple mechanical method?"

If Dr. Foster had used Thomas's, Hewitt's, or Jackson's pessary, I think he would have been more successful in Case 3.

The following cases were reported by myself in the March number of the *American Homœopathist*, 1878:

"CASE I.—A young, healthy married woman, whom I had treated a year previous for retroversion, became pregnant, and about the *fifth* week was attacked with violent and almost constant vomiting. Her mother and friends dissuaded her from calling a physician, because 'the vomiting belonged to pregnancy, and there was nothing to be done.' About the seventh week, however, the husband became anxious and called me in. From my previous knowledge of her history, I immediately suggested the probability of a retroflexion. She was loath to have an *examination*, and her mother 'never heard of such a thing.' I therefore gave Nux vomica, which seemed indicated; but in a few days I was summoned again. I then made an *examination*, and found the uterus retroflexed. Dr. Thomas's retroversion pessary was placed in position, which changed the retroflexion into the normal position of the uterus. After this she had no vomiting or nausea, except the *ordinary morning sickness*. I italicize the last phrase in order to call attention to the fact that in the majority of cases, where the vomiting or nausea *grows worse as the day advances, we may safely diagnosticate some uterine displacement.*"

"CASE II.—Mrs. B., multipara, pregnant with her third child. Four years ago she miscarried at three months, 'from excessive vomiting,' as her medical attendant said. She was now about two months advanced, and vomited everything she ate or drank. The empty retching was very agonizing. When I was called it was for *dysenteric symptoms*, and here I wish to make another point, namely, *that when a pregnant woman, before the fifth month, complains of tenesmus, with small mucous or watery evacuations, look for retroversion of the uterus*. In this case I insisted on an *examination*, and found not only a retroverted but a very sensitive and inflamed uterus. It was with difficulty replaced, and required a long Jackson's retroversion pessary to retain it in its proper position. The next day all but the morning nausea ceased, but it required absolute rest, with side and face position in bed for several days before the pain, soreness, and rectal irritation disappeared. The use of Arnica and Caulophyllum proved valuable."

"CASE III was one of a more unfortunate character. Mrs. C.,

married a few months, 'passed over her time' ten days, when she was suddenly attacked with violent vomiting, occurring from walking or riding, or when expelling a constipated stool. I gave her Nux v., Cocculus, and Argentum, which corresponded with the three above given causes; but no benefit followed their use. I had treated her before her marriage for retroflexion, and I now suspected the same condition. An *examination* proved the correctness of my opinion. The uterus was pushed up, an Albert Smith pessary placed under it, and the vomiting ceased. She wore the pessary two weeks, then at her request I removed it. For ten days all went well, when, after a long walk and climbing many stairs, the vomiting again occurred, and continued two days, or until a pessary was inserted. Again at the third month the pessary was removed, and notwithstanding the persistent use of Sepia and Lillium, the uterus became retroflexed at *three and a half months*, this time from straining at stool, she thought. Again the pessary was inserted with prompt arrest of the vomiting, which was this time so violent as to lead to spasmodic retching and vomiting of bloody mucus. She wore this last pessary two weeks, when, contrary to my advice, she removed it herself. I then cautioned her very impressively against any severe exercise, such as going rapidly upstairs or down, riding over rough roads, or straining at stool. For about two weeks she was perfectly free from any pain or discomfort, and everything bid fair for a perfectly normal advance of pregnancy. At this time I visited the Centennial Exposition, and then went abroad with my family. A few days after my departure, she entertained at her house some friends from California, and unfortunately felt obliged to attend them to places of amusement, and on long rides over the rough wooden pavements of Chicago. On one of these rides, her husband informed me that she was attacked with such severe pain in the *back* and lower bowels, with extreme nausea, that he was obliged to place her on a street car and take her to his residence. After arriving home she was attacked with violent retching and vomiting of watery mucus mixed with blood. A prominent homœopathic surgeon of this city, now deceased, was called in, but from some inexplicable reason did not recognize the cause of the trouble, notwithstanding the woman's suggestion as to the nature of the previous attacks, and the means I had taken to remove them. She rapidly grew worse, the pain in the back and hypogastrium became agonizing, the vomiting became grumous, and delirium supervened. At this juncture, the attending surgeon, for an equally inexplicable reason, transferred the case to a neighboring physician, whose utter ignorance of the pathological condition of the uterus, and the cause of the attack, was manifested by his absurd symptomatic treatment. In consequence of the total neglect of the ordinary and simple treatment of this case, the poor victim died on the fourth day, after expelling in her last agonies a fœtus of nearly five months.

"These cases, especially the last, should serve as practical lessons to all physicians, not to overlook the displacements of the pregnant uterus, but to act promptly and decidedly, and not to rely upon medicines to remove a condition only amenable to replacement and mechanical support."

Out of several cases of the above character which have come under my observation and treatment, since the first edition of this work, I select two typical ones. The important lesson they convey is (1) that a large proportion of the profession are inexcusably ignorant of the frequency and grave consequences of the displacement, and (2) that any method of treatment except the mechanical is worse than useless, because it is malpractice.

CASE I.—Mrs. F., a Swede, pregnant for the first time after marriage of several years. Was not aware that she had any uterine trouble previous to marriage or since. When about seven weeks advanced in pregnancy was seized with violent pains in the back and hypogastrium; frequent urging to urinate; constipation, with constant urging to stool; violent, empty, retching, etc. A neighboring physician (homœopath) was called in who treated the case for a week without producing the slightest amelioration, then becoming alarmed he recommended that she consult some other physician. She called me, and after hearing the description of her sufferings I demanded an examination. The uterus was found completely *retroverted*—the fundus against the lower rectum, the os pressing against the bladder.

With considerable difficulty, owing to the extreme sensitiveness of the uterus, which was on the verge of inflammation, I replaced it *in situ*, and placed a Jackson retroversion pessary under it. In less than an hour—she informed me the next morning—nearly all her sufferings left her, only a dull soreness remained. I then wrote a note to her former attending physician, describing the character of the displacement (which he had not suspected) and stating the means I had used to prevent inflammation and miscarriage. I also advised that unless the pessary became displaced not to remove it for a month or two. But if it became displaced to place it in proper position. About a week after, while making a bed, she felt something give way in the pelvis, and was seized with the pain, etc., above described. Dr. —— being called, and shown my letter, said "he did not believe in any mechanical supports for the uterus," and insisted upon removing it altogether, which he did then and there. Immediately her sufferings became aggravated and attained in a few hours an intense degree of severity. Labor-like pains set in, accompanied by violent, empty retching and ineffectual tenesmus. Becoming alarmed, the husband sent for me, but being absent in Texas, my associate, Dr. Newman, who had been informed of the nature of the case, visited her, and gave her almost immediate relief by replacing the uterus and placing the pessary in position to support it. In a few days she was on her feet, and at this writing she is now four and a half months advanced. The pessary has been removed some weeks, and she is in perfect health.

CASE II.—Mrs. E., a fine-looking young German woman, married a year, was seized, when six weeks advanced in pregnancy, with violent retching and vomiting, excruciating pains in the stomach, hypogastrium, sacral region and hips. She had been troubled with *retroversion* before conception, and had been properly treated for it by an intelligent eclectic physician, wearing at that time an "Albert Smith pessary." But when called to see her at this attack, diagnosed "inflammation of the stomach," and treated her accordingly, not demanding an examination, or suspecting anything present but an aggravated case of the usual "vomiting of pregnancy," tending to inflammation. The patient becoming alarmed by the violence of her sufferings, demanded a change of physicians, and I was sent for. After getting a history of the case I felt that all other treatment but mechanical would be unavailing. On examination the uterus was found *flexed* and retroverted—it was almost crescentic in shape. Carefully elevating it upon the promontory of the sacrum, I placed a Jackson pessary under it, and ordered her to lie on the face and side, (she had formerly been ordered to lie altogether on the back.) *She did not vomit or retch once* after the uterus was replaced, and for the next few days experienced only the usual slight morning nausea. On the third day she had a constipated stool, during which she strained hard to expel it, when she felt as if the pessary had got out of place, and shortly after, the former symptoms returned, but not as severely as at first. On my arrival I found that the pessary had *tilted to one side*, owing to its soft, yielding character, and the fact that the lower portion was not long enough to rest *between the ramus of the pubis*. It is important to select a pessary that when placed in the vagina, will *not* rest on the pubic arch, but project slightly *under* the arch. The point should be felt just behind the urethra. This allows the pessary to move slightly downward when coughing, vomiting, during defecation, and even when taking a deep inspiration. I have repeatedly observed that a pessary properly fitted will move slightly upward and downward with the movements of respiration. This method of placing a pessary avoids any shock to the uterus or urethra, during coughing, etc.

Selecting a Thomas' improved retroversion pessary, it was placed in the position above described, and almost immediate relief followed. The next day the woman was on her feet, and after the usual morning nausea, spitting, etc., was able to eat three light meals a day. Several weeks have passed and she is in excellent health and spirits and will not hear to having the "support" removed. She is quite right, for if a pessary is well fitted I never knew it to cause irritation in such cases, but will absolutely prevent miscarriage and displacement, and should be worn until the end of the third month. Some of my cases have worn them until the time of "quickening," for retroversion may and has occured as late as the fifth month, and even later.

For general directions relating to the mechanical treatment of *flexions* and *versions* of the uterus, by the use of pessaries, refer to the chapter on that subject. I will only add, that if the proper pessary is selected, and properly placed, so that it does not cause irritation by being too large, too small, or of improper shape, there is not the least danger of inducing miscarriage, but, on the contrary, it will do more to prevent that accident than any other means at our command.

The *medicinal treatment* of the nausea and vomiting of pregnancy may be arranged in three classes:

(1.) Those homœopathic to the flexion and congestion of the uterus.

(2.) Those homœopathic to a localized gastric irritability.

(3.) Those homœopathic to abnormal reflex irritation.

To the *first* class belong *Aletris, Belladonna, Cimicifuga, Nux, Lilium, Sepia, Apis, Pulsatilla.*

To the *second* class belong *Antimonium, Ipecac., Iris,* Nux, Kreosote, Oxalate of Cerium, Pepsin, Bismuth, Columbo, Veratrum, Pulsatilla, Sanguinaria, and Mercurius sol.

To the *third* class belong the *Bromides of Potassa, Ammonia,* and *Camphor;* also, *Chloral hydrate, Cocculus, Calabar,* and *Jaborandi.* Electricity has relieved many cases where I have advised it, when all medicines failed. For the special indications for the above remedies I refer the reader to the *Materia Medica* and special works on Obstetrics.

I will, however, add my own experience with some of the medicines I have mentioned.

I have met with cases of vomiting during pregnancy, so severe that the life of the woman was almost despaired of; cases where not a drop of water or a particle of food could be tolerated by the stomach, and where the uterus appeared to be in a normal condition. In these cases there may have been abnormal conditions of the uterus which could not be ascertained, such as fibrous tumors, presence of exudative products, tumefactions of the uterine tissues, inflammations of the placenta, decidua, and other membranes. In these cases all the ordinary remedies proved unavailing. But I have never failed to carry my patient through, although I was obliged to resort to procedures which are not put down in our textbooks, namely:

In four cases I kept the women in the horizontal posture for several weeks, giving meanwhile from 20 to 40 grains of *Bromide of potassa* twice a day, in an enema of a pint of milk, beef tea, or mutton broth, allowing nothing whatever to be swallowed, thus giving the stomach perfect rest. I had every reason to be gratified by the result of this treatment, for in a few weeks they became conscious of an ability to retain food, and the trial was successful. In many cases, two or three enemas of Bromide of potassa, 20 grains at night, has completely arrested the violent morning vomiting. The same results have often followed the administration of 15 grains in an ounce or two of milk, before rising in the morning. I have used *Chloral hydrate* in the same manner, and with excellent success, being led to its use by seeing its good effects in sea-sickness. These remedies do not cure by acting on the uterus or stomach, but by lessening and even arresting altogether the hyperæsthetic condition of the reflex nervous system, giving the stomach rest until it recovers its tone.

Associated with the vomiting of pregnancy, there is often a condition which may be designated as "extreme biliousness." The tongue is foul, coated with a brown, dirty fur, the breath is fetid, the complexion dirty-yellow, and constipation is obstinate. In this condition no remedy will relieve as well as *Merc. sol.*, *Hahn.*, two or three grains of the 2^x, or even 1^x, given every three or four hours until the tongue cleans and the bowels are opened. *Podophyllin* 2^x, or *Eupnomin* 1^x, will often do as well, and should be used if the woman has ever been salivated, or cannot bear mercurial preparations.

On several occasions I have verified both Hahnemann's and Ringer's recommendation of the value of *Ipecac.* in obstinate and *continuous* vomiting during pregnancy. I have cured cases with the 6th dil., and have made equally good cures with five-drop doses of the mother tincture, or the wine of Ipecac. Kreosote and Carbolic acid in minute doses have helped me out in some obstinate cases, while in those cases where *undigested food* was invariably vomited, I have seen the best results from *Pepsin*, in the form of saccharated powder, wine of pepsin, or lacto-peptin, a dose given before each meal, even if the meals were taken every few hours. *Bismuth sub. nit.*, in doses varying from $\frac{1}{100}$ to 5 grains of the crude, taken before or after food was taken, often acts magically in removing oppressive pain, eructations, and vomiting of food. *Ferrum*, especially the *lactate*, in grain doses of the 1^x, is often useful in similar cases in pale, anæmic patients. *Digitalis* 1^x will often give surprisingly good results when, in persons of slow, languid, irregular circulation, they complain of palpitation, sinking at the stomach, vertigo, faintness, etc. Koumiss often arrests the vomiting of pregnancy, and at the same time affords excellent nutrition.

I cured two cases with *Cuprum acet.*, when with the vomiting there were cramps in the stomach and suffocative paroxysms. In another case, where the womb was very painful and sensitive, a *Belladonna* plaster on the hypogastrium arrested both the pain and the vomiting.*

In several cases, where there was the most obstinate and uncontrollable constipation, it, with the vomiting, were both removed by 5 drops of tincture Calabar bean, given at night for a week. In another case I was led to the use of Jaborandi 2^x, by the excessive salivation which annoyed the patient. That medicine removed in a week both the salivation and the distressing nausea.

The *treatment* of those cases of vomiting which occur after the seventh month of gestation, depends on the nature of the cause, which is supposed to depend on constriction of the fibres of the os and cervix uteri. These instances are not usually continuous with the vomiting of the earlier months, but occur after a long interval of quiet.

In one case reported within a few years, the vomiting was so violent and exhaustive that a sponge-tent was introduced for the purpose of inducing premature labor, when, to the astonishment of the physician, its removal the next day was followed, not by labor, but by a complete cessation of the vomit-

---

* Cazeaux (*Obstetrics*) recommends us often successful, painting the cervix uteri with Ung. bell., or applying a tampon moistened with the diluted extract.

ing. In another case of vomiting in the ninth month, the attending accoucheur attempted forcible dilatation of the os with his fingers, in the hope that premature labor would occur, but instead a permanent cure of the vomiting resulted. Since these cases were published, several reports have appeared in American and foreign journals, confirming the value of gentle dilatations of the os uteri in this late vomiting. One such case occurred last year in my own practice. All that is necessary is to carefully insert the forefinger in the os, and press gently in all directions, so as to slightly relax the constricting circular fibres.

It is possible that some of the medicines known to exert a relaxing effect on these muscular fibres, might be used successfully for this form of vomiting. *Belladonna*, internally and topically; *Stibium*, internally; *Gelsemium*, *Cimicifuga* and *Cannabis indica*. Recent observations seem to prove that *Morphia* has this effect; it is said to relax the os when its rigidity causes slow and retarded labor, and it might be given in minute doses for the vomiting, for it is certainly very homœopathic to vomiting of a reflex or secondary character.

Some singular facts have come to my knowledge relating to the little understood and complex relations of the uterus to other organs, and the exceptional great tolerance of the uterus under certain circumstances.

(1.) During the meeting of the National Medical Association (allopathic) in Chicago, in 1878, Dr. M. O. Jones of that city informed Dr. J. Marion Sims that he had for years arrested the obstinate vomiting of pregnancy by placing a piece of lunar caustic, half an inch long, in the cervical canal, allowing it to remain there fifteen or twenty minutes. Dr. Sims seemed much interested in the relation, but made no comments at that time. But afterwards, while in Europe, he published an account of the procedure in the *London Lancet*, and stated that he had tried the method in many cases with success, and *without causing miscarriage*. This would seem to show that there was a *point of irritation* somewhere in the cervical canal that caused the vomiting of pregnancy, and perhaps the vomiting which sometimes occurs in non-pregnant women with uterine disease. The supposition is that the "charring" of the cervical mucous membrane produced a "sedative" effect on its nerves, and arrested the reflex irritation. It would seem to show that *flexion* was not always the cause of the vomiting, but some other irritating factor. Of course miscarriage *would* be caused if the caustic was pushed through the *inner* os. I cannot recommend the practice, for I have never tried it. It should not be forgotten, however, in those cases which seem to threaten life.

(2.) During the meeting of the Illinois State Society in 1878, Dr. G. C. McDermott, a delegate from Milwaukee, informed me that Dr. Ford of that city recently had a severe case of intractable vomiting in a pregnant woman, who had suffered similarly in three previous pregnancies, in one of which miscarriage was brought on, as it was deemed necessary to save her life, at least such was the decision of a consultation of eminent physicians. In this fourth pregnancy a similar decision was reached after serious consultation. A uterine sound was used, but could not be introduced; a sponge tent was introduced, but doubtless only part way up to the inner os. *This arrested the vomiting* for a week, and on its return another tent was used with like effect, and this procedure was repeated several times. In this case the amelioration was not permanent; but if a flexion was the cause, a retroversion pessary might have continued the palliation to a permanent improvement. It is possible that the tent acted like the lunar caustic, and benumbed the sensitive point in the cervix.

These cases are suggestive, and should lead to more careful study of the singular anomalies of uterine disturbance.

ALBUMINURIA.

By *albuminuria*, it is implied that through the medium of the kidneys the albumen is filtered off from the blood to a greater or less extent, and discharged from the system in the urine. When this occurs it is believed, also, that the kidneys fail in some measure to depurate the blood by eliminating urea. Albuminuria and uræmia are not identical terms, as either condition may exist and the other be absent; but I suppose that albuminuria cannot be developed to any considerable extent without being accompanied by more or less uræmia (Barker).

Albuminuria of pregnancy is *not* Bright's disease, for the albumen of Bright's disease differs essentially from that occurring in the temporary albuminuria of pregnancy, as can easily be shown by its chemical reactions. The albumen of the urine in Bright's disease when brought in contact with the Oxide of copper assumes a beautiful reddish-violet color, and produces a more or less abundant flocculent black precipitate. The urinary albumen of pregnancy, when Bright's disease does not exist, while it coagulates readily by heat and Nitric acid, does not exhibit any such reaction with the Oxide of copper. Robin has demonstrated that granular casts are not characteristic of any particular morbid state or pathological change of structure of the kidneys.

It is an accepted fact that in a large number of cases, gestation develops a temporary albuminuria which may disappear during or soon after puerperal convalescence. The phemomena pertaining to this condition are rarely manifested before the *sixth* month of pregnancy. Statistics seem to prove that it occurs more frequently in first than in subsequent pregnancies. It has been supposed that there is some close relation between albuminuria and convulsions during labor; but when careful and repeated examinations of the urine during gestation have failed to detect albumen in the urine, convulsions have occurred during labor, and *afterwards* the urine has been found loaded with albumen. It is now settled that *mechanical* obstruction of the circulation during pregnancy, while it may be *one* cause of albuminuria, is not always the cause. A sudden cold may develop the disease. The presence of albumen in the urine has been regarded as the *cause* of many of the pathological conditions occurring during pregnancy and the puerperal state, when, in reality, it was only an *effect*.

Albuminuria when present during pregnancy usually causes: (1.) *Headache*, severe and persistent, usually associated with sleeplessness, impaired vision, hesitation or embarrassment in speech, and great nervous irritability. It may culminate in delirium, coma, paralysis, hemiplegia and convulsions. (2.) *Œdema*, first of the face, worse in the morning; then of the lower extremities and even general anasarca. (3.) *Gastric irritability*; sometimes obstinate constipation, and sometimes diarrhœa. (4.) *Miscarriage* or *premature labor*.

Treatment.—There are three important indications for treatment, namely:

I. *To relieve the hyperæmia or congestion of the kidneys.*

II. *To prevent the impoverishment of the blood which results from the loss of albumen.*

III. *To prevent the various disturbances which often terminate in paralysis or convulsions.*

Before proceeding to the special therapeutics of albuminuria, I deem it my duty to enter my serious protest against the manner in which it is treated by our best homœopathic text-books. In Guernsey's *Obstetrics*, the etiology of the disease is very fairly given, but when it comes to that all-important point, the treatment, what do we find? A careful examination into the merits of medicines having a specific action on the kidneys? Medicines capable of curing the symptoms and pathological conditions accompanying the disease? *Not at all!* Nothing but three pages of a "repertory," or an enumeration of medicines causing such symptoms as relate to color, sediment, smell, and painful sensations in the kidneys, bladder, and urethra. Nine-tenths of the medicines mentioned are not renal remedies at all, and the symptoms they cover have very little, if any relation to the renal functions, or the abnormal condition of the urine found in albuminuria. He refers us to another section entitled "Urinary Difficulties," in which thirty-three medicines are mentioned, and their supposed characteristics given. But out of the thirty-three, only ten at most have any specific action on the kidneys or the urine. And this is called the correct homœopathic treatment of albuminuria! To make matters worse, the senseless plan of enumerating the medicines in alphabetical order is retained, instead of giving preference to those which experience has found most efficacious. Further comment on such so-called "treatment" is unnecessary. It is like the "apples of Sodom"—

"Fair to the eye,
But turn to ashes on the lip."

I do not claim for the following treatment anything infallible for our knowledge of the specific treatment of albuminuria is still in its infancy. But I do claim that it has a basis of common sense, and has in many hands besides my own proved efficacious in many cases.

I. *To relieve the hyperæmic or congested kidneys.*

If the woman is *plethoric*, and there is persistent redness of the face, injection of the conjunctiva, hot skin, sharp pains in the head, a hard labored pulse, denoting arterial tension, and an elevation of temperature, there are three medicines which are capable of removing the hyperæmia by lessening the abnormal blood-pressure. In these cases, the condition of the urine is of less moment than the threatening constitutional, pathological state.

Veratrum viride by its secondary symptoms corresponds exactly to this general hyperæmic condition. In the experiments narrated in the "*Therapeutics of New Remedies*," after the first stage of depression had passed, or even when a toxic dose was given, the arterial blood-pressure rose enormously, and every organ became congested—even the convulsive centre in the medulla—and violent convulsions set in. According to my *law of dose*, which I believe to be as important as our *law of cure*, Veratrum viride should be prescribed in the lowest dilutions, or the crude preparation. I have succeeded best by giving one to five drops of the tincture or 1^x every two hours, until the pulse, temperature, and other symptoms showed that the arterial tension was normal. It will not do to suspend the medicine at this point, but continue it at longer intervals, in order to hold the advantage gained, and alternate with it some other remedy corresponding to the special symptoms of the urine, etc.

Gelsemium corresponds to nearly the same general condition, but the pulse is not as *large, hard*, and *heavy*, nor is the urine as *scanty* as under Verat. vir. The dose is, however, nearly the same. There is moreover a tendency to stupor, and a dimness of vision not found in Verat. vir.

Aconite is indicated when with a *small, hard, wiry* pulse, there is *great restlessness and anxiety, fear of death, scanty urine*, and acute rheumatic-like pains all over the body. The dose will vary from five drops of the 1^x to the same quantity of the 2^x dilution.

I have made no mention of albumen in the urine or dropsy as indicating these medicines, because they are only indicated in the first stage of the disease, or for dangerous conditions of arterial excitement, which may arise in the later stages.

If the constitutional symptoms are not of a nature to indicate the above three arterial sedatives; if there is not general increased blood-pressure or congestion to the brain, but we believe the kidneys to be highly congested, with only sufficient intracranial hyperæmia to cause frightful dreams, incoherency of speech, temporary attacks of delirium, etc., another class of remedies becomes necessary.

These remedies should have a specific influence, not only on the kidneys, but the capillary circulation, in general. They need not be *primarily* homœopathic to *renal* hyperæmia, but to general hyperæmia. One of the most important of this class, I believe to be the *Bromide of lithia*. Its action in albuminuria, and the conditions tending thereto, is as follows: (1.) It relieves the renal hyperæmia by contracting the arterioles of the kidneys. The blood-pressure in those vessels being lessened, the urine increases in quantity, and the elimination of urea is increased. It thus prevents an accumulation of this toxic agent in the blood, and the tendency to uræmia is lessened.

The action of this medicine can be favored and increased by the use of certain mineral waters, namely, Gettysburg, Bethesda, Vichy, or Seltzer; if these are not readily obtainable, pure soft water, filtered rain-water, or a weak solution of Bitartrate of potassa, or Citrate of potassa. No active diuretics should be used in this stage of the disease, for they all tend (in material doses) to increase the renal hyperæmia. (In some cases the *Benzoate of lithia* may be substituted, namely, when the appearance of the urine corresponds to the indications found under Benzoic acid).

The Bromide of lithia controls admirably the cerebral irritation, the rheumatic pains, and many of the abnormal mental phenomena. The *dose* which I have used and advised most successfully, is *five grains* three or four times a day, and if obstinate insomnia occurs, with nocturnal frightful dreams, or delirium, I give an additional dose of fifteen or twenty grains at bedtime. For this last purpose, however, the Bromides of lime, soda, or potassa may be used.

Another class of remedies, more important for the radical *cure* of albuminuria, are those medicines which are *primarily* homœopathic to the pathological condition existing in the kidneys, viz.: *Turpentine, Cantharides, Apis mel., Copaiva, Erigeron, Equisetum, Nitrate of uranium, Phytolacca, Eupatorium purp., Helonias, Aurum muriaticum, Merc. corrosivus, and Arsenic*. These are the most important, and it is rare that any others will be found more than palliative, because their action is not powerful enough in the direction of the real pathological state of the kidneys.

Turpentine is probably the closest *similimum* to both acute and chronic albuminuria. *Primarily* it causes the *acute* stage, *secondarily*, the *chronic*. But the disease must be idiopathic, *i. e.*, not originated by, or kept up by disease of the heart. In the attenuations, from the 3^x upward to the 30th, it is primarily indicated in active congestion of the capillaries, the Malpighian tufts, and the glomeruli, with exudation of albumen. It also destroys and expels the glandular surface of the tubuli uriniferi. The urine contains blood, epithelium, cylindrical casts, and coagula, and coagulates under the action of heat and nitric acid. Its ultimate secondary effects are complete denudation of the surface of the tubuli, of all of its glandular cells, paralysis of its nervous supply, passive hæmorrhage, uræmia, and convulsions.

When the stage of acute congestion, or inflammation of the kidneys subsides, and structural changes set in, the higher dilutions of Turpentine are useless. When pus, disorganized blood-corpuscles, fat globules, and very scanty urine, loaded with albumen, appears, then Turpentine should be given in the lowest dilutions.

The action of *Erigeron* and *Erechthites* is closely analogous to that of Turpentine, and may sometimes be substituted for that drug.

Cantharides, next to Turpentine, is the closest *similimum* of acute congestive albuminuria. When the patient seems suffering from some feverish erethism, with scanty and frequent urinations, irritation of the generative organs, burning or scalding pains during micturition, and pain in the back, Cantharis 3d or 6th will ward off impending albuminuria, or arrest the disease even after albumen has appeared in the urine. It is of little value after œdema has appeared, for it causes not so much structural changes in the glandular elements of the kidneys as paralysis of its nervous supply. If homœopathic to dropsy at all, it is a dropsy from renal paralysis. If homœopathic to uræmia, it is from the same cause. In allopathic practice it has warded off uræmic convulsions with suppression of urine when given in material doses. The material doses acted curatively because the paralytic condition was similar to the secondary action of Cantharis.

Apis has many symptoms in common with Cantharis. It causes both congestion and paralytic conditions in the kidneys with cerebral hyperæmia. But it differs from Cantharis in its specific power of causing œdema of almost every organ, and effusion of serum into almost every cavity. Very soon after the renal irritation of Apis commences we find symptoms of general dropsical effusion. Even œdema of the brain and lungs sets in. Feverish symptoms are rarely present, but the blood is soon poisoned, and this causes a condition known as uræmic fever, which finds relief in eruptions on the skin (urticaria, erysipelas, etc.). In this respect it closely resembles the preparations of Arsenic.

Copaiva is a medicine of much greater value in albuminuria than is generally supposed. It is especially indicated when the origin of the disease is a cold or sudden check of perspiration, and the affection of the uriniferous tubes are *catarrhal* as well as congestive. It has caused acute attacks of albuminuria, with feverish symptoms, bloody urine, and all the symptoms of acute congestion of the kidneys. In the middle attenuations it will effectually remove these conditions. It has been found singularly efficacious in dropsy after all other remedies had failed, and when the urine was almost entirely suppressed; but in these cases it was prescribed in material doses (five to fifteen drops of the balsam repeated every few hours). The dropsy of Copaiva is a secondary effect of that drug, due to conditions of the kidneys somewhat similar to those caused by Cantharis.

Equisetum is recommended by Dr. Marsden as having proved curative in many cases of "painful dysuria with albuminous urine in pregnant women," but our knowledge of its pathological effects is quite limited. (See Pathogenesis in Allen's *Encyclopedia of Materia Medica*.)

Nitrate of uranium seems indicated where an attack of congestive albuminuria has set in after diabetic symptoms.

Helonias is indicated in the same condition, but it is a much more useful remedy on account of its specific relation to the uterus. Albuminuria is supposed by some to be due to reflex irritations transmitted from the uterus to the kidneys. Helonias may be useful when this variety exists. But it is also homœopathic to acute albuminuria, for it has caused that condition when taken in large doses. It is best indicated when the blood is impoverished by the drain of albumen, diabetes, or from failure of nutritive processes.

Eupat. purp. closely resembles Helonias, but will be found more useful when general œdema is present. Here the lowest dilutions or the tincture will have to be used in order to restore the normal quantity of the urinary secretion.

Mercurius cor. has been found one of the most valuable and trustworthy remedies we possess when albumen appears in great abundance in the urine of pregnant women. Under the use of the 3^x trit. the albumen has rapidly disappeared, and serious results been prevented. But I must warn the physician not to continue Merc. cor. but a few days in this low trituration, as it may cause serious general disturbances and a final aggravation of the albuminuria. I would advise that as soon as improvement fairly set in to change the 3^x for the 3^c, or the 6^x trituration.

Aurum, Cuprum, Arsenicum, and *Argentum* are excellent remedies, but will be considered in the next section. I ought to add that in all acute cases where congestion of the kidneys is evidenced by *pain and tenderness* in that region, and the urine is *scanty*, "*smoky*," *and high-colored*, that great relief may be obtained, and the action of specific remedies favored, by the application of *dry cups* over the lumbar region, or a poultice of flaxseed meal, in which is mixed a spoonful of tincture of aconite root. Hot "*sitz*" baths act in the same manner, by attracting the blood to the surface capillary bloodvessels, and thus lessening the amount of blood in the kidneys.

II. *To prevent the impoverishment of the blood which results from albuminuria.*

Next in value to these medicines, which are strictly homœopathic to the morbid condition existing in the kidneys, are those remedies which arrest the chloro-anæmia from loss of albumen. The physician must not confound the *hydræmia* which often exists with the anæmia, with true plethora. Hydræmia is a kind of serous plethora, which closely simulates true plethora, causing great disturbances of the circulation and even serious local congestions.

The most valuable remedy for this condition of the blood is *Ferrum*. Iron is purely homœopathic to many forms of anæmia, chloro-anæmia, and even hydræmia, but it cures by its secondary action, for the true plethora caused by large doses of iron is often followed by unmistakable anæmia. It is very important that we select the proper preparation of Ferrum. According to all trustworthy observers the *Ferrum muriaticum* is the best. It not only acts by increasing the red globules, but it also acts on the kidneys, increasing the watery portion of the urine.

The large doses of Tincture of chloride of iron prescribed by the opposite schools, have doubtless done great injury by causing primary pathogenetic effects, which secondarily aggravated the chloro-anæmia. The 2^x or 3^x dilution of Tinc. ferr. mur. in 10-drop doses after each meal is amply sufficient for all curative purposes. Next in value, I consider the *Albuminate of iron*, a few drops of the officinal syrup given in the same manner.

The trophic spinal nerves which influence nutrition have

much to do in the production of anæmia. The condition of the blood in albuminuria tends to paralyze these nerves. When Ferrum alone does not seem to act favorably, *Ferrum et Strychnia citras* 1^x will be found an admirable remedy. If there is deficient digestion of food, vomiting of ingesta, lientery, give *Pepsin* or *Lacto-peptin*, 5 grains after each meal. If actual hydræmia exists, with general œdema, irregular circulation, and congestive symptoms, Digitalis is a powerful curative remedy. The action of the heart in this condition is one of excessive action, irregular, but with insufficient radical power. Digitalis, by imparting regular and normal force to that organ, prevents the congestive phenomena, and increases the favorable effect of restorative medicines. Five to ten drops of the 1^x dilution may be given before each meal, and Ferrum after. A favorite preparation of mine in hydræmia with weak and irritable heart is *Digitalin* 3^x and *Ferrum met.* 1^x, equal parts, triturated together, of which I prescribe two or three grains three times a day. Should there exist an unnaturally profuse flow of the urine with anæmia, instead of scanty flow, I have found *Lycopus*, 5 or 10 drops four times a day, to have a better effect than Digitalis in regulating the irritable heart.

There are other medicines which should be consulted in these cases, medicines which are eminently restorative in their effects on the blood and the processes of nutrition. Among the best are *Aletris*, *Helonias*, *Calc. hypophos.*, *Viburnum prun.*, *China*, *Hydrastis*, *Phosphoric acid*, and *Kali chlor*. They have also another action, especially useful in hydræmic albuminuria and anæmia, namely: to *prevent miscarriage, premature labor*, or *the death of the fœtus*. Nor should we forget in such cases to resort to *Cauloph.*, *Cimicifuga*, *Mitchella*.

Gallic acid should not be forgotten when we are selecting medicines to prevent the chloro-anæmia due to loss of albumen. It is not only powerfully palliative in albuminuria, but may prove curative in some cases. No medicine so surely prevents the loss of albumen after scarlatina, and it also arrests the loss of blood from the kidneys which sometimes occurs as a sequel to that disease.

If, during the obstinate albuminuria of pregnancy, other

medicines fail to arrest the loss, and the woman begins to look pale, bloated, and anæmic, give her 5 to 10 grains of the 1^x trituration of *Gallic acid*, watching the urine to observe if the percentage of albumen is decreasing. Ferrum can be given at the same time, the former before, the latter after meals. So soon as the waste of albumen is arrested, change from Gallic acid to some more radically curative remedy.

III. *To prevent the nervous disturbances which often terminate in paralysis or convulsions.*

Medicines alone cannot be relied upon for this purpose. We should advise and insist that all emotional excitement should be avoided, and all overtaxing the physical powers in any way. The digestive organs should be carefully watched; the diet so regulated as to avoid indigestion, and, above all, the patient should not become constipated. We should see that the woman sleeps and lives in apartments which are well ventilated and free from sewer-gas, or air impregnated with carbonic oxide, for these two agents poison the blood in a most dangerous manner, and serve to precipitate an attack of uræmia.

If the woman shows signs of a mania in which *fear of death* is the prominent symptom, give *Aconite* 6th or 30th, or *Arsenicum* 6th or 30th. If a *suicidal* mania obtains, no remedy is so efficient as *Aurum*, especially Aurum muriaticum, or the "Muriate of gold and soda," as both these latter salts have potent diuretic and eliminant properties, and are useful in dropsy or uræmia. They should be prescribed in about the 3d trituration.

A mania with *melancholy* is usually controlled by *Cimicifuga*, *Helonias*, or *Hellebore*. If violent *delirium* with *insomnia* occurs, it is of little use to select the medicine by the *ensemble* of its mere symptoms. We must select that medicine which will have the twofold action of calming the irritated brain and rapidly eliminate the noxious poison in the blood. I have already mentioned the value of the *Bromide of lithia*, and I cannot too strongly recommend that it be used freely (twenty to thirty grains every four or six hours) until the delirium is controlled.

The obstinate *sleeplessness*, which sometimes precedes the mania, can be controlled by Scutellarin 1^x, Cypripedin 1^x, or Caffein cit. 2^x. (A new preparation, the Bromohydrate of Caffein, in the 2^x trit., has acted promptly in a few cases in my practice.)

There is a group of symptoms and conditions denoting *uræmic poisoning*, which may occur, and which demand the promptest action if we would save life, namely, *coma, paralysis, general dropsy with suppression of urine*, and *constipation*. In this condition I have several times adopted the following treatment with success. Mix two to four drops of *Croton oil* with a little butter, and force it back upon the posterior portion of the tongue. It soon dissolves and gets into the stomach, causing very soon profuse watery discharges from the bowels, which relieves the brain, nervous system, and kidneys, and allows us to select the proper specific remedy.

In less grave cases, where the above condition is impending, and the danger is imminent, the patient's life has been saved by the timely administration of $\frac{1}{8}$ or $\frac{1}{10}$ of a grain of *Extract of elaterium*. This dose causes in four or five hours very profuse watery evacuations from the bowels, sometimes a quart at a time, followed by profuse diuresis. Not only in the uræmia of pregnancy have I warded off coma and convulsions with this medicine, but I believe I have saved life in the extreme dropsical conditions which occur during the progress of organic diseases of the heart. The above dose can be repeated three or four times during twenty-four hours, until the patient is relieved.

If we fail, as in rare cases we may, in causing proper evacuations from the bowels or kidneys, we have in the new remedy, *Jaborandi*, a drug which for its action on the skin and salivary glands has no equal. Fifteen to thirty drops of a good tincture, or a few grains of the 1^x trit. of its active principle, the *Muriate of pilocarpin*, will cause such profuse sweating that the thickest bed-clothing will be saturated, and such profuse flow of saliva that it has been known to amount to twenty ounces in a few hours. Cases have been reported where, in dropsy with coma or convulsions after scarlatina, or puerperal uræmia, the action of *Jaborandi* has doubtless saved life by its rapid elimination of morbid matters and water through the skin.

CONVULSIONS.—If puerperal convulsions occur, even when albuminuria has not been suspected, or from evident uræmia, there are but few remedies which can control them. It is useless to waste time in selecting from the list given in our text-books. (Of the thirty-five mentioned by Guernsey only three are of any value, Gels., Verat. vir. and Bell.)

Veratrum viride, when we have the violent tonic convulsions, tetanic, with turgid head and face, pulse large and hard and bounding. Give large doses, five to ten drops every half hour, until the spasms are arrested, and the pulse softened. In *New Remedies* many cases of fearful severity are reported cured by doses as large as twenty to forty drops, repeated in an hour or two if necessary. So long as the intense blood-pressure exists in the arterial system, even such large doses are safe.

Gelsemium ranks next, but corresponds to a milder form of spasms, with less pressure in the great arteries, but more in the capillaries, a softer pulse, redder face, and suppressed urine.

Chloral hydrate has been used very successfully in *ante-* and *post-*partum convulsions. The dose is twenty to thirty grains, repeated in two hours if necessary. (See *New Remedies*.)

Bromide of lithia (or any bromide) when the convulsion has the characteristics of epilepsy or hysteria.

Cuprum acet. 2^x, when the characteristic of the spasm is the terrible suffocative phenomena, and the "beginning in the fingers and toes."

Belladonna, Agaricus, Cicuta, Solanum, Hyoscyamus, and *Stramonium* may each be valuable in cases where these symptoms closely correspond to those of the convulsive disorder, but I cannot recommend them in cases where the blood-poisoning is evidently the cause, while I believe they may be promptly efficacious when the convulsions arise from some other excitement of the nerve-centres than uræmia.

CONSTIPATION is a far more important abnormal condition than many writers would have us suppose. Many women whose alvine evacuations are perfectly normal and regular when not pregnant, become very constipated so soon as *conception* occurs. This fact would go to prove that constipation is caused by some undiscovered cause, and not from mechanical obstruction. It has generally been supposed that it was due to the pressure of the growing uterus on the upper part of the rectum. It is thus explained by Guernsey: "The mechanical pressure exerted on the rectum, by which its calibre is diminished and its action paralyzed, and the habits of inactivity in which some pregnant women indulge, especially in cities, combine to produce costiveness, and as a final result of the constipation, hæmorrhoids, either blind or bleeding, appear in many cases. And the very great amount of vital force consumed in the womb may also tend to draw away from the intestinal canal some of the energy that might have sustained its regular and daily evacuations."

This last sentence explains the cause of many cases of constipation in pregnant women, and should be borne in mind in our treatment. When constipation occurs at or after the sixth week, I believe it is due in many instances to retroversion, for I have found the uterus retroverted in such cases, and the constipation was removed promptly by replacing the uterus and keeping it in proper position by a well-selected pessary. But constipation may *cause* retroversion both by the accumulation of fecal matter above the uterus, and by the efforts in straining at stool which some women improperly make. Besides this result, constipation may cause headache, anxiety, giddiness, sleeplessness, distressing dreams, vomiting, fissure of the anus, swelling of the veins of the legs, tedious labor, irregular and deficient pains, obstruction to the passage of the child, and puerperal fever. It may even produce febrile excitement, loss of appetite, erratic pains in the bowels, simulating false pains. The efforts of straining to relieve the bowels may also result in abortion. I have met with cases where the pains were so severe that it seemed as if retroversion or miscarriage must occur, but an examination would reveal a rectum enormously enlarged and obliterating the vagina, requiring copious and repeated enemas, and even the use of small placental forceps or the fingers to remove the hardened masses. Campbell had a case in which the bowels were so overloaded that after the birth of the child the attendant thought the woman had another child to bear. The rectum was found distended to the size of a quart bottle, and the woman died of inflammation of the bowels. Fourteen pints of fecal matter were removed after death from the small bowels, after the colon and rectum had been relieved during life. Churchill once attended a labor in which the hollow of the sacrum was nearly filled up with a hard mass; but a more careful examination proved it to be the lower bowel filled up with hardened fæces, giving to the finger the sensation of a large growth upon the bone. Great difficulty was experienced in emptying the bowels, and not until then did labor progress favorable. Ashwell mentions many similar cases.

Treatment.—It is evident that the constipation of pregnant women should receive judicious and effectual treatment if we wish to avoid many of the sufferings of that period, and also difficult and painful labor. The treatment is to be based on somewhat different principles than would guide us in prescribing for ordinary cases of constipation.

When called upon to prescribe for a case, we should inquire into the diet, habits of life, and previous state of health. We should advise active exercise, walking is better than riding; also regulate the diet by advising coarse bread, fruit, etc. If these means fail we may then select the medicine which seems appropriate. Out of the fifty-five medicines recommended by Guernsey, only eleven are of much value, namely: *Alumina, Bryonia, Ignatia, Lycopodium, Nux vom., Opium, Plumbum, Ratanhia, Sepia, Sulphur,* and *Verat. alb.* Of these I have succeeded best with Alumina 30th, Bryonia θ in five-drop doses, Ignatia 30th, Lycop. 6th trit., Nux vom. 2^x, Opium 12th, Plumbum 2^x, Ratanhia 3d, Sepia 6th, Sulphur 1^x trit. (See special indications in any text-book.)

But there are other remedies which I value more highly.

Calabar bean (Physostigma) is very efficient when we have a lax, distended state of the intestines, with large quantities of flatulence, fæces dry, hard, and very large. Five drops of the 1^x dil. four times a day will promptly remove the inactivity, by contracting the half-paralyzed muscular coats and restoring peristaltic action. I have seen better results in some cases from giving ten drops of the crude tincture at night, than from smaller doses repeated during the day.

Strychnia in the 6th trit. will often act favorably when Nux or Ignatia fails.

Opium 2^x trit. will often restore the action of the bowels when the higher attenuations do not act.

Podophyllin is one of our most efficient medicines, but it must be given in appreciable doses, for it secondarily causes constipation. It is indicated by the constantly furred tongue (yellow or brown coating), the sallow complexion, anorexia, and a tendency to piles, fissure, or prolapsus recti. I first try the 2^x trit., three or four doses a day. If this fails give granules, each containing the ½ or even ¼ grain, once, twice, or thrice daily, until free laxative effects occur; then lessen the dose for a few days, when it will be found that the bowels will remain regular. The alternation, or even combination of Podophyllin with Physostigma or Nux vomica, will often give better effects than either alone. There will now and then occur cases so obstinate that all the above remedies will fail, and we feel that we must use some palliative measures. This fact has time and time again been recognized by the best men of our school, and even the most bigoted high-dilutionists have been *forced* to use them. I cannot too highly commend Dr. Madden's excellent article on this subject in the *British Journal of Homœopathy*, vol. vii, page 310.

In such intractable cases when the bowels *must* be moved occasionally, we should select those laxative and aperient agents which produce the least irritation, and do not run counter to the idiosyncrasies of our patient.

I have generally succeeded with the Podophyllin above mentioned. But some women cannot take it in sufficient doses without "griping." A pill of Aloes, 1 gr.; Ext. Hyos., ½ gr.; Soap. 2 grs., combined (one at night), is often very easy and effectual, but they should not be too long continued.

A powder composed of Sulphur, Senna and Liquorice root (from a formula found in the Prussian Pharmacopœia), is a great favorite with many physicians and patients. It is prescribed as " Comp. Glycyrrhiza powder—30 to 60 grs. in a wineglass of water at night." It usually gives one or two free evacuations the following morning.

In some cases in plethoric women, with rush of blood to the head, feverishness, abdominal congestion, etc., a wine-glass of Hungarian or German bitter water, taken in the morning before breakfast, acts very favorably, especially if a glass of common water be drank afterwards.

In very acid states of the stomach, with sick headache, a teaspoonful of Tarrant's Seltzer Aperient before breakfast is very efficient.

Some non-medicinal substances will often act as excellent palliatives, namely: (1.) A tablespoonful of wheat *bran*, mixed with milk and taken at night or morning. (2.) An equal quantity of flaxseed or white mustard seed, taken in the same manner. (3.) A dozen stewed prunes, taken at night on going to bed, or in the morning before breakfast. (4.) A teaspoonful of tamarinds mixed with a wineglass of water, taken in the same manner. (5.) The eating of any ripe fruit before breakfast—oranges, bananas, apples, pears, berries, etc.

I trust that the importance of inquiring into the condition of the bowels during the few days or hours preceding labor will not be forgotten.

Nature usually attends to that matter by causing considerable relaxation, with several free discharges the day before, or just before the commencement of labor. Women should be told of the importance of an open state of the bowels, and advised to use enemas to bring about that result if nature proves delinquent. The first question that the physician should ask on entering the lying-in room and examining his patient, should relate to the state of the bowels and bladder. If these have not been evacuated, it should be done immediately, else he may have to conduct a painful, protracted, and otherwise disagreeable labor.

CHAPTER III.

MEDICATION OF THE FŒTUS IN UTERO, CONSIDERED AS A MEANS OF PREVENTING DYSTOCIA.

ONE of the most important questions now under discussion by the medical professions is, "Can we influence the fœtus by administering medicine to the mother?"

The literature of this subject is exceedingly meagre. The regular text-books on obstetrics merely give the physiological theories of the placental circulation or gaseous interchange; not one of them (except Richardson's) refers to the transmission of medicinal agents to the child through the maternal blood. Schroeder alone mentions the experiments of Reitz, who, after injecting Cinnabar into the blood of a pregnant rabbit, found the red particles of that chemical in the blood of the fœtus particularly distinct in the capillaries of the pia mater. Schroeder therefore considers the transmigrations of maternal blood-cells into the blood of the fœtus easily conceivable. The older journals contain only investigations on the normal placental respiration, which leaves the question more or less undecided. In later journals, all German, occur a few papers on this subject. Gusserow injected Tincture of iodine and a solution of Ferrocyanide of potassium into the stomach of pregnant rabbits, guinea pigs, and dogs, but was unable to find any traces of the drugs in the liquor amnii and in the urine of the fœtus, even when the dose had been given five days before. On the other hand, he was able to detect Iodine in the liquor amnii and in the urine of the newborn infant after administering Iodide of potash for some time (about two weeks) to the mother before her delivery.

Zwisfel, in 1874, claimed that in five cases where Chloroform had been inhaled during labor, traces of it were found in the placenta and in the urine of the child. He claims that the fœtus respires through the placenta, and that "this respiration is subject to the same conditions as that of the animal after birth."

Benicke reported to the German Medical Association, in 1875, that he gave Salicylic acid to twenty-five women during labor, and found it in the urine of the children immediately after birth, the shortest time after its administration being forty minutes. Ruge and Martin report precisely the same observation with Salicylic acid; but Iodide of potash was found by them only in small quantities after prolonged use by the mother. Fehling failed to get any effects on the fœtus by giving Woorara to the mother rabbit. He got no effect on fœtal rabbits by giving Chloroform to the mother.

Fehling and Korman's observations seem to show that the hypodermic injection of Morphine during labor does affect injuriously the fœtus before and after birth. In a number of the *American Journal of Obstetrics*, for 1877, this subject of poisoning the fœtus by giving Opium or Morphine to the mother, was discussed at length by many of the most prominent obstetricans of this country. Dr. Paul F. Mundè said: "With reference to the influence of medicinal agents on the infant during pregnancy, there was not much to be said. We all know that by giving the mother tonics and various nutritious medicinal agents, we aid in securing a vigorous and healthy offspring; we also know that by putting a syphilitic mother under specific treatment during her pregnancy, we are preserving the child from premature death, or for a time, at least, from venereal disease. But still we do not know why the fœtus in utero is not poisoned by a drug, given to the mother in a dose, adapted to her, it is true, but large enough to be fatal to the child after birth."

Dr. Fordyce Barker concluded his paper by laying down the following proposition as his belief: "1. There is no evidence which can be accepted in science that *narcotic* drugs administered to the mother ever produce their specific effects on the fœtus *in utero*." He says nothing about *other* medicines, however.

Dr. Peaslee reasons from a very material ground, namely:

because the fœtus having no *vascular* connection with the mother, derives no part from her of its blood—therefore: mere solutions in the mother's blood of medicinal substances are not absorbed into the fœtal blood in any *appreciable* amount. His argument, however, disproves nothing, for he would say the same of the administration of attenuated medicines, and we know that in this his argument would be fallacious.

Dr. Mundè, after listening to this celebrated discussion in the New York Obstetrical Society, said: "It matters not if in one hundred experiments the substance injected into the maternal blood is *not* discovered in that of the fœtus in ninety-nine, if it be so discovered in the hundredth, it shows that such a thing is *possible*.

The homœopathist will bear in mind that in all these experiments and investigations *crude* medicines were used. Now we know that attenuated or material medicines do not depend altogether upon their mixture with the blood for their action on the organism. Many medicines act as do emotions, dynamically. We all know that mental influences affecting the mother during pregnancy or lactation, do certainly affect the fœtus in utero or the infant at the breast. Surely the blood has nothing to do with the influence in such cases.

I have watched closely the various papers and investigations, as well as the discussions going on in the journals and societies of the allopathic school during the last few years, and I have no hesitation in expressing the opinion that there is no reasonable doubt that medicines of all kinds do affect the child in utero when given to the mother. I think also that we can entertain no doubt as to the specific action of homœopathic medicines on the fœtus in utero when administered to the mother. This leads to a discussion of

Dystocia due to the Fœtus.

In the previous chapters we have considered only those causes of dystocia which belonged to the maternal organism; but there is a dystocia which is due to the foetus, and the causes may be classified as follows:

UNNATURAL OSSIFICATION OF THE SKULL.—A variety of this condition is that known as complete ossification. Dr. Joulin describes one kind as "the development of ossa Wormiana in the fontanelle, causing solidification." But all the sutures may be united by ossification to such an extent that the head is unyielding, and has to be forced through the genital canals without undergoing any change. It is probable that this condition may be amenable to treatment, by so arranging the diet of the mother as to leave out all or nearly all the earthy salts which go to make bony tissue. (This will be treated of in another place.)

Another cause of dystocia is

HYDROCEPHALUS OF THE FŒTUS.—The hydrocephalic head sometimes reaches an enormous size. Cazeaux says Meckel has a skull of an infant whose transverse diameter is 16½ inches, and its height from the occipital foramen to the vertex, 16 inches. Burns gives a case of hydrocephalus where the circumference was 23 inches.

The management of a labor obstructed by a hydrocephalic head is not within the scope of this work. It is possible that we may be able to prevent this abnormal condition by medicating the mother. If the mother has been delivered of one hydrocephalic fœtus, she may give birth to another. If we have reason to fear this, the treatment should consist in placing the mother upon a diet of animal food almost exclusively, a diet the opposite of that advised for too rapid and complete ossification. We may also administer to the mother such remedies as *Sulphur*, *Calcarea*, and *Phos.* (as recommended by Grauvogl), also *Apis* and *Arsenicum*, in the highest potencies, or *Kali hyd.*, *Merc. iod.*, and *Calc. iod.*, in the lowest attenuations, or in nearly a crude state. The treatment of ascites or hydrothorax in the fœtus should consist of the administration of the same remedies.

TUMORS OF THE FŒTUS.—Large tumors may form upon the body, or in the body of the fœtus. They may attain such a size as to prevent delivery without mutilation of the child or mother. It is possible that knowing the kind of tumors affecting a previous fœtus we may give remedies to the mother which may prevent their recurrence. Tumors are sometimes hereditary, and when the mother is affected with them it is possible that the persevering use of appropriate remedies would prevent their appearance in the fœtus.

MONSTROSITIES.—The various anomalies which have been described under this head are usually the result of two general causes, namely: (1.) Some psoric, scrofulous, or strumous taint in the blood of the parents, and (2.) Mental emotions, shocks, etc., to which the mother may have been subjected. Homœopathic remedies have such a decided action over mental aberrations and emotional shocks, that there is good reason to suppose that if *Aconite* is given to a pregnant woman immediately after a *fright*, it might prevent an injury to the fœtus in utero. For the effect of any of the injurious mental shocks, select the appropriate remedy recommended in our Repertories, and we may succeed in arresting deficient or abnormal development.

Dr. Croserio, now dead, and a distinguished physician of our school, wrote a small work entitled *Homœopathic Obstetrics*, the first of that kind which appeared in our literature. He quotes "a distinguished physician" (name not given) as having "just published a very interesting memoir upon this subject," in which he recommends giving the mother, at different periods of pregnancy, and at long intervals, Sulph. 30th, and Calc. carb. 30th, to *purify the fœtus* from the psoric (scrofulous) taint which it may have inherited from its parents. Several homœopathic physicians claim to have seen good results from such a procedure.

But Sulphur and Calcarea are not the only remedies to be prescribed. The whole history of both parents should be obtained, and, after a careful comparison of all the symptoms, the specific remedy should be fixed upon, and given at least once a week all through the pregnant state. Some mothers invariably give birth to *rickety* children. To such should be given *Silica*, *Calc. phos.*, or *Calc. hypophos.*, in the triturations from the 3d to the 30th.

When tuberculosis is feared, give the mother Calc., Silica, Kali carb., Iodine, etc.

The Hypophosphites of lime, soda, and potassa, in small doses, using glycerin as a vehicle, are admirable prophylactics of tuberculosis.

But of all remedies *Oleum jecoris* is probably the best, for it contains all of the above remedies in an attenuated form, and is more readily assimilated in the oil than in any other way. It is not necessary to give massive and nauseous doses. The 1^x or even 3^x trituration is efficacious, a few grains three times a day. If, however, the mother was greatly emaciated, and her stomach would readily tolerate it, I would prescribe a teaspoonful three times a day.

It is not best to rely on medicine alone. Advise the mother to remain much in the pure open air, and live on an appropriate diet, else all remedies may fail of preventing disease in the offspring.

HARDNESS OF THE BONES OF THE FŒTUS.—One of the chief obstacles to easy and natural labor is often the unnatural size, the strength, and unyielding character of the osseous structure of the child. A child weighing *ten* pounds can be delivered easily through a natural pelvis and soft parts, if the bones of the head are soft and yielding. The head readily elongates and becomes smaller in diameter, assuming a wedge or cone shape, permitting its easy expulsion. But if the bones are very unyielding, a long, tedious, and painful labor results, with more or less injury to the mother and child, even if the forceps are skillfully used.

It is said that this condition of the fœtus can be remedied and prevented by placing the mother upon a diet composed of such articles as contain the *least amount of bone-making materials*, namely: earthy salts, lime, silica, common salt, etc. This diet is sometimes called the fruit diet, although other articles than fruit may enter into the diet of the mother. Drs. Lewis, Cummins, and Richardson have called attention to this matter, and allege good results from its use. The following is the history of the discovery of the method, and the details of the method itself and its results.

Fruit Diet.—In 1841, there was privately printed in England, a small pamphlet of twenty-two pages, in which a gentleman, who was a chemist, gave an account of an experiment he himself tried in the case of his wife, whose labors had been so excessively painful that there was much reason to fear that she would not survive the next one. The result was so favorable that he felt it his duty to publish it, with his name and residence.

A few experiments were made in Boston and vicinity with distinguished success, when the discovery of Ether rather threw it into the shade. As, however, there are persons, especially out of New England, who do not use Ether, the following extracts are made from the pamphlet in question, which has now become very scarce, and indeed practically inaccessible. It will be best to begin by stating the principle of the system, with which the experimenter ends his account. In proportion as a woman subsists during pregnancy upon aliment which is free from earthy and bony matter, will she avoid pain and danger in delivery; hence the more ripe fruit, acid fruit in particular, and the less of other kinds of food, but particularly of bread or pastry of any kind, is consumed, the less will be the danger and sufferings of childbirth.

"The subject of this experiment had, within three years, given birth to two children, and not only suffered extremely in the parturitions, but for two or three months previous to delivery her general health was very indifferent; her lower extremities exceedingly enlarged and painful; the veins so full and prominent as to be almost bursting; in fact to prevent such a catastrophe, bandages had to be applied, and for the few last weeks of gestation her size and weight were such as to prevent her attending to her usual duties. She had on this occasion, two years and a half after her last delivery, advanced full seven months in pregnancy before she commenced the experiment, at her husband's earnest instance; her legs and feet were as before, considerably, swollen, the veins distended and knotty, and her health diminishing. She commenced by eating an apple and an orange the first thing in the morning and again at night. This was continued for about four days, when she took just before breakfast, in addition to the apple and orange, the juice of a lemon mixed with sugar, and at breakfast two or three roasted apples, taking a very small quantity of her usual food, viz., wheaten bread and butter. During the forenoon she took an apple or two and an orange. For dinner she took fish or flesh in a small quantity, and potatoes, greens, and apples, the apples sometimes peeled and cut into pieces, sometimes boiled along with the potatoes, sometimes roasted before the fire and afterward mixed with sugar. In the afternoon she sucked an orange or ate an apple or some grapes, and always took some lemon-juice mixed with sugar or treacle. At first the fruits acted strongly on the stomach and intestines, but this soon ceased, and she could take several lemons without inconvenience. For supper she had again roasted apples or a few oranges, and rice or sago boiled in milk; sometimes the apples peeled and cored were boiled with the sago. On several occasions she took for supper apples and raisins, or figs with an orange cut among them, and sometimes all stewed together. Two or three times a week she took a teaspoonful of a mixture made of the juice of two oranges, one lemon, half a pound of grapes, and a quarter of a pound of sugar or treacle. The sugar or treacle served mainly to cover the taste of the acids, but all saccharine matter is very nutritious. The object in giving these acids was to dissolve as much as possible the earthy or bony matter which she had taken with her food in the first seven months of her pregnancy. She continued in this course for six weeks, when much to her surprise and satisfaction the swelled and prominent state of her veins, which existed before she began, had entirely subsided; her legs and feet, which were also swollen considerably, had returned to their former state, and she became so light and active she could run up a flight of more than twenty steps with more ease than usual when she was perfectly well. Her health became unwontedly excellent, and scarcely an ache or a pain affected her up to the night of her delivery. Even her breasts, which at the time she commenced the experiment, as well as during her former pregnancies, were sore and tender, became entirely free from pain, and remained in the very best condition after her delivery also, and during her nursing.

"At nine o'clock on the evening of March 3d, after having cleaned her apartments, she was in the adjoining yard shaking her own carpets, which she did with as much ease as any one else could have done. At half-past ten she said she believed her 'time was come,' and the accoucheur was sent for. At one o'clock the surgeon had left the room. He knew nothing of the experiments being made, but on being asked, on paper, by the husband, two days afterward, if he 'could pronounce it as safe and as easy a delivery as he generally met with?' he replied, on paper, 'I hereby testify that I attended Mrs. Rowbotham on the 3d inst., and that she had a safe labor, and more easy than I generally meet with.' On his asking the female midwife if she thought it as easy as usual, she replied, 'Why, I should say that a more easy labor I never witnessed. I never saw such a thing, and I have been at a great many labors in my time.'

"The child, a boy, was finely proportioned and exceedingly soft, '*his bones being all in gristle*,' but he became of large size and very graceful, athletic, and strong as he grew up. The diet of his mother was immediately changed on his birth, and she ate bread and milk and all articles of food in which phosphate of lime is to be found, and which had been left out before. She also got up from her confinement immediately and well. Mr. Rowbotham made a table of substances, with the proportion of phosphate of lime in each, so that it may be avoided in the food during pregnancy, and used afterward in nursing, when the bones and teeth of the child are made.

"Beans, rye, oats, and barley, *have not so much earthy matters* as wheat. Potatoes and peas not more than *half* as *much;* flesh of fowls and young animals *one-tenth;* rice, sago, fish, eggs, etc., *still less;* cheese, *one-twentieth;* cabbage, savoy, brocolli, artichokes, colewarts, asparagus, endives, rhubarb, cauliflower, celery, and fresh vegetables generally, *one-fifteenth;* turnips, carrots, onions, radishes, garlic, parsley, spinage, small salad, lettuce, cucumbers, leeks, beet-root, parsnips, mangelwurzel, mushrooms, vegetable marrows, and all kinds of herbs and flowers, average less than *one-fifth;* apples, pears, plums, cherries, strawberries, gooseberries, raspberries, cranberries, blackberries, huckleberries, currants, melons, olives, peaches, apricots, pineapples, nectarines, pomegranates, dates, prunes, raisins, figs, lemons, limes, oranges, and grapes, on the average are *two hundred times* less ossifying than bread or anything prepared from wheaten flour. Some articles, as honey, treacle, sugar, butter, oil, vinegar, and alcohol, if unadulterated, are quite free from earthy matter. But still worse than wheaten flour is common salt, and nearly as bad are pepper, cinnamon, nutmeg, cloves, ginger, coffee, cocoa, Turkey rhubarb, licorice, lentils, cinchona, or Peruvian barks, cascarilla, sarsaparilla, and gentian.

"With regard to drinks, no water except rain and snow, as it falls, and distilled water, is free from earthy matter, and every family should have a distilling apparatus; and perhaps it would pay capitalists to form a company for the purpose of distilling water on a large scale. Filtering water is not sufficient to purify it of earthy matter, because a filter can only remove such particles as are mechanically mixed.

"An American lady, who usually suffered terribly in labor, immediately procured the pamphlet and governed her diet by it partially, and had the easiest labor she ever had. Another who governed herself *wholly* by it from the first moment she was aware of being pregnant, like the English lady never experienced a moment's discomfort before delivery. She had taken nothing made of our grains, but confined herself to the best Indian ones, rice, sago, tapioca; and taking a disgust to our summer fruits, subsisted largely on oranges, tamarinds, marmalades, and also took a great many lemons. At first the fruits made her bowels too loose, but she did not abandon them on that account, but took mutton broth with rice in it to correct this effect. She also took fish and sardines, and the young of meats; for the older animals are the greater quantity of earthy matter is contained in their secretions, and so it is even with milk. She had so little thirst that she drank nothing but a little tea made with distilled water. This lady and her husband were neither of them very young, she was thirty-five and he forty at the birth of her eldest child; and she had been an invalid in her chamber from fifteen to thirty years of her life, though very well at the time of her pregnancy, and for the first time in her life taking much exercise in the open air. Consequently, and because of her extreme nervous delicacy, she did not escape pain in the labor the first time, and the process was several hours. But in the two succeeding times, at the last of which she was forty, the labors were very short and not at all severe. In all the cases she rigidly adhered to the diet without a single day's exception, and her three children were perfectly splendid instances of large, healthy, strong, and beautiful *physique*. The youngest of them is now eighteen years of age."

A common error is, that during gestation the mother needs to "eat for two;" that is, that more food is necessary to support properly herself and her growing infant than at other times. This is a thorough delusion. On this point, and on diet during pregnancy generally, Dr. Bull, a very sensible and experienced English physician, says:

"We habitually take more food than is strictly required for the demands of the body; we therefore daily make more blood than is usually wanted for its support. A superfluity amply sufficient for the nourishment of the child is thus furnished, for a very small quantity is requisite without the mother on the one hand feeling the demand to be oppressive, and on the other, without a freer indulgence of food being necessary to provide it. Nature herself corroborates this opinion; indeed, she solicits a reduction in the quantity of support rather than asks an increase of it; for almost the very first evidence of pregnancy is the morning sickness, which would seem to declare that the system requires reduction rather than increase, or why should this subduing process be instituted? The consequences, too, which inevitably follow the free indulgence of a capricious, and what will afterward grow into a voracious appetite, decidedly favor this opinion; for the severest and most trying cases of indigestion are by these means induced, the general health of the female disturbed and more or less impaired, and through it the growth and vigor of the child. . . . .

"If the appetite in the earlier months, from the presence of morning sickness, is variable and capricious, let her not be persuaded to humor and feed its waywardness from the belief that it is necessary to do so; for if she does, she may depend upon it, from such indulgence it will soon require a larger and more amply supply than is compatible with her own health or that of her little one.

"If the general health before pregnancy was delicate and feeble, and, as a consequence of this state, it becomes invigorated and the powers of digestion increase, a larger supply of nourishment is demanded, and may be met in such case without fear; for instead of being injurious, it will be useful.

"Lastly, a woman, toward the conclusion of pregnancy, should be particularly careful not to eat in the proportion of two persons, for it may not only bring on vomiting, heartburn, constipation, etc., but will contribute, from the accumulation of impurities in the lower bowel, to the difficulties of labor."

A few figures given by Dr. Dewees, whose discussion of this subject is exactly in harmony with Dr. Bull's, show very clearly the absurdity of the idea that it is necessary to "eat for two." They are in substance as follows:

"On an average, a newborn child, together with all the accompanying materials expelled at birth, weighs not more than ten pounds, viz.: eight pounds for the child itself, and two pounds for the placenta, etc. A table of 7077 births in Paris gave an average of about two pounds less than this, being for the child itself just six pounds. Now, a daily supply of less than three-quarters of an ounce, during the average of two hundred and eighty days of pregnancy, will amount to this ten pounds; and this daily supply is decidedly less than the average quantity of unnecessary food which is usually eaten. Since, therefore, we almost always eat too much, and since the the ordinary overplus is more than enough to supply the requirements of pregnancy, and particularly since the natural symptoms of that state usually indicate less food rather than more, it is mere common sense to conclude that pregnant women neither *want* nor *need* to 'eat for two.' The fact is more likely to be the seeming paradox that enough for one is too much for two; *i.e.*, that less food than usual, rather than more, is best during pregnancy."

Regularity in hours of eating is advantageous to the health, and more care even than usual should be taken during pregnancy to observe this practice. Another, almost, or quite equally important rule is, to eat nothing for four hours, or at least for three hours before going to bed.

Eating should also be, as indeed it should always be, in moderation. It should be deliberate, and it should be cheerful. Deliberation is almost indispensable to moderation; for it is the sense of satisfaction of hunger that tells us when to stop eating, and this sense is blunted and almost useless when the food is swallowed rapidly and without thorough chewing; and the appetizing effect and healthful stimulus of cheerfulness at meals are too well known to require any detailed enforcement in this place.

"I once heard a physician object to the fruit diet on the ground that it would not give enough strength to the pregnant woman to undergo the severe trial of parturition. But if the trial be robbed of its severity, where is the need of the extra strength? Again, what strength is imparted by rich strong animal food, that is ejected from the stomach, or else passes out of the system undigested and unassimilated? Or, if this food be digested and assimilated, what is the result? Why, plethora, that most dreaded foe of the pregnant woman, leading either to miscarriage or to a terrible confinement, in which the chances of death are greatly augmented."—(*Dr. Cummings.—"Richardson's Obstetrics."*)

Dr. Verdi says, in his advice to mothers:* "It will do no harm to avoid what is repugnant to you, but it may be detrimental to your health to satisfy the longing for slate-pencil, chalk, or other deleterious substances which sometimes women in your condition crave.

"But above all, keep a cheerful mind and do not yield to grief, jealousy, hatred, discontent, or any perversion of disposition. It is true that your very condition makes you more sensitive and irritable; still, knowing this, control your feelings with all your moral strength.

---

* On Maternity. By T. S. Verdi, M.D.

"If you believe that strong impressions upon the mother's mind may communicate themselves to the fœtus, producing marks, deformity, etc., how much more should you believe that irritability, anger, repinings, and spiritual disorders, may be impressed upon your child's moral and mental nature, rendering it weakly or nervous, passionate or morose, or in some sad way a reproduction of your own evil feelings; and, indeed, this is more frequently found to be the case than is the physical marking of a child by its mother's impressions.

"If a woman needs culture and expansion, both of her perceptions and conceptions of the beautiful, in order to produce a grand poem or painting and sculpture, or to conceive noble measures for the relief of the sufferings of others, then does she also need all these for that highest of all her efforts, when it seems as every fibre of her being was put upon the stretch to do its share in the grand donation to love and to humanity of a child."

SUMMARY.—It appears then that painless parturition may be secured by attention to the following points during pregnancy (besides correct previous bringing up, moral and mental, and physical):

Moderate healthful exercise and avoidance of shocks, fatique, and overexertion.

Comfortable or at least quiet and patient mental condition, avoiding all bad tempers.

Amusement and agreeable occupation as far as possible.

Judicious use of bathing, particularly of the sitz-bath.

The fruit diet, and avoidance of unsuitable food, and of alcoholic, narcotic, and other stimulants.

Watchfulness and prompt treatment of the various ailments of the situation, should they appear.

Cheerfulness on the part of the patient, and kindness and indulgence by the husband and friends.—(*Dr. Cummings.*)

CHAPTER IV.

IMMEDIATE TREATMENT OF FUNCTIONAL DYSTOCIA.

By immediate treatment is meant the application of medicinal agents which have the power to control or modify those conditions and symptoms which may arise during the progress of labor, and tend to produce an abnormal manifestation of that condition.

EXTREME SLOWNESS OF THE LABOR.—A duration of eighteen or twenty hours in a primipara cannot be regarded as an alarming circumstance. But if labor is prolonged beyond this period, some assistance is demanded. Assistance may be given before, however, if certain symptoms arise. The *first* stage of labor, that of dilatation of the cervix, may be prolonged without danger; but the *second* cannot pass beyond certain limits without greatly endangering the health of the patient, or the life of the child. It is found that the latter is lost at least one time in four, when the head remains in the excavation longer than seven or eight hours after the complete dilatation of the os uteri and the rupture of the bag of water; whilst it nearly always survives when the first period is prolonged to forty, fifty, or even sixty hours. (*Cazeaux.*)

The first stage of labor, even prolonged as above, rarely presents any serious symptoms, except great fatigue, nervous irritation, loss of sleep, depression of spirits, and alarm. These symptoms may be met successfully by proper food, gruels, and broths, and the use of *Coffea, Cimicifuga, Ignatia, Aconite,* and *Erythroxylon coca.*

It is during the first stage of labor that we meet with those two conditions of the *os* and *cervix,* which cause great trouble and anxiety, namely:

1. *Rigidity of the cervix.*
2. *Spasmodic contraction of the cervix.*

Rigidity is a passive force by which the fibres of the neck of the uterus resist the dilatation they have to undergo.

Spasmodic contraction is an active force by which the fibres contract and diminish the size of the opening, previously exhibited by the mouth of the womb.

In *rigidity* the tissues seem dense and like a piece of leather soaked in grease. The labor continues without dilatation of the orifice, which retains a certain thickness against which contractions strive in vain, until the woman is exhausted with her fruitless efforts.

Pain in the loins, according to Madame Lachapelle, is a diagnostic sign of rigidity of the os.

Spasmodic contraction may occur after the cervix has attained considerable dilatation.

The orifice presents a thin cutting edge, and is warmer, drier, and more sensitive to the fingers, and very irritable. This extreme sensibility—tenderness—of the neck is often the only symptom by which we can decide that we have a spasmodic contraction to deal with.

These two conditions must not be confounded with a neck which continues thick, simply because the contractions are *insufficient*, badly directed, or lost against some mechanical obstacle in the pelvis.

The obstetric authors of our school have always advised the same remedies for both conditions. Nothing could be more unscientific or irrational, for the conditions are opposite.

In *rigidity* the true remedies are—*Gelsemium, Lobelia, Veratrum viride, Passiflora* (or Curare), and *Nux vomica*.

Gelsemium, when the face is flushed, the woman is plethoric, dull, and apathetic, the pains irregular in force and frequency, and not in their proper place—passing from before backward, or occupying one side of the uterus—and the os is thick, sodden, but unyielding. The efficient dose is a drop or two of the tincture or 1^x every half hour until the cervix relaxes. (Dr. Page asserts that it acts better when administered in *hot* water.)

Lobelia is next in value. It is indicated for the thick, leathery, unyielding cervix, and the reflex symptoms which arise from the obstruction—the dyspnœa and nausea.

In some rare cases this medicine will have to be given until it causes general relaxation, but no vomiting need be induced. However, I think I should not hesitate to give it until vomiting ensued if the rigidity did not yield, for I should only be imitating nature. In such cases vomiting often occurs spontaneously, with immediate relaxation of a rigid os. *Dose* usually the same as Gelsemium.

Stibium, in doses of $\frac{1}{60}$ of a grain, or a few grains of the 2^x trit., will often act better than Lobelia, while the indications are nearly identical.

Veratrum viride is useful in those cases where the old physicians considered "bleeding" absolutely necessary, *i. e.*, when the woman was very plethoric, the head and chest congested, the pulse full and bounding, and eclampsia threatening. It should be given in doses of one to five drops of Norwood's tincture, repeated every hour. Usually three or four doses bring down the great blood-pressure, the congestion, and the cervix relaxes normally.

Passiflora or *Curare* are indicated if the rigidity seems almost tetanic, with a general tendency to tetanic stiffness, such as we will sometimes meet with in hysterial subjects. The dose is, of the former, ten to twenty drops of the tincture every hour; of the latter, a few drops of the 1^x or 2^x dilution.

Nux vomica is also indicated in the same condition, especially if the woman feels the peculiar pains in the "loins" and a constant urging to stool. *Ignatia* may better suit some women.

These two last remedies being *primarily* indicated should be prescribed in the tenth or thirtieth attenuations. A single dose is often sufficient.

I have induced the needful relaxation in some cases without the use of medicine, by placing the woman in a hot (100° F.) *sitz-bath* for half an hour or more, and injecting, at the same time, warm water against the os.

I think I have observed that the internal administration of *the* remedy, particularly Gelsemium, was aided by applying it topically. Saturate a tampon of cotton or soft sponge, with a warm solution of glycerin and water, equal parts, and pour upon it 15 or 20 drops of the medicine. Apply this to the rigid *os*, and wallo it to remain.

Spasmodic contraction of the os and cervix requires the following remedies:

Aconite, Amyl, Belladonna, Hyoscyamus, Solanum, Cactus, Lachesis, Conium, Caulophyllum, Cimicifuga, Morphia, and *Viburnum.* (Also *Ether* and *Chloroform.*) *

Aconite, when the patient has the usual restlessness, anxiety, and fear of death, some fever, with fine, small hard pulse, the vagina and os, dry, hot and sensitive. Here the 3^x or 6^x acts promptly—especially so if we apply it topically, as above recommended.

Amyl nitrite.—Although no cases are on record where this agent has been used for the specific purpose of relaxing a spasmodically constricted os, it has been used by some English and Continental physicians in "very painful labors," and with alleged good results. The most painful of all labors are those in which there is spasmodic constriction of the os, with spasmodic contractions of the uterine body.

The pathological condition is similar to that which obtains in angina pectoris, or dysmenorrhœa from constricted cervix. The value of Amyl in both these conditions has been established. It should prove equally useful in dystocia from spasm of the os. A few drops, not more than *five*, should be poured into a small vial, on a little cotton, or upon a handkerchief, and the patient be directed to inhale deeply eight or ten times, or until the face flushes and the head throbs. When this occurs, the constriction of the os will be observed to relax, and allow the expulsive efforts of the womb to accomplish their purpose. Instead of inhalation, the Amyl may be given internally, a few drops of the 2^x dilution every ten minutes.

Belladonna† has been alternately praised and denounced as a

---

* The special indications, and *contra*-indications, for the use of *Ether* and *Chloroform*, are to be found in all text-books on *Obstetrics*. They frequently relax spasmodic rigidity of the cervix and os, when internal medicines fail.

† *Atropin in Spastic Rigidity of the Cervix in the First Stage of Labor.*—In the July Number of the *American Journal of Obstetrics* (1878), appeared a paper by H. L. Horton, M.D., in which he strongly advocates the use of atropin when the os is spasmodically rigid, and the pains fail to cause any normal relaxation. He applies it by means of a hypodermic syringe, having a needle hooked at the end. After hooking the index finger of the right hand into the anterior lip of the

remedy in "rigid os." This conflict of opinion is explained by Cazeaux, who writes: "The Belladonna, so highly lauded by some accouchers, is by others thought to be useless. It seems to me that this difference of opinion has arisen from confounding *simple rigidity* with *spasmodic contraction*. Though without action in the former case, I think it very useful in the latter."

We have the same difference of opinion in our school, often from the same cause, but generally on account of a want of understanding of the relation of *dose* to primary and secondary effects of medicine. In order to prescribe Belladonna with any approach to scientific precision or curative effect, we must know something as to its method of action; which of its effects are primary and which secondary. The primary action of Belladonna (also Conium, Hyoscyamus, Stramonium, and Solanum) is to *relax and paralyze the sphincters of the orifices of hollow organs*, but not the muscles of the organs themselves. Its secondary effects are just the opposite, the *sphincters are affected with spasm and constriction*, and the muscular fibres of the walls of the organs themselves, weakened.

Now according to the only *law of dose* known, Belladonna should be given in a high potency when we have symptoms simulating its primary effects, and in low attenuations when the symptoms resemble its secondary effects. In cases of spasmodic constriction of the cervix, Belladonna is secondarily indicated, and no possible benefit can arise from its administration in highly attenuated doses.* The low dilutions must be

---

cervix, and drawing it slightly forward, the needle is carried along the palmar surface of the finger, keeping the point firmly pressed against it, so as to avoid wounding the maternal parts. After carrying the point of the needle within the cervix, it is raised from the finger and buried by slight traction somewhat deeply into the muscular structure of that portion of the uterus. After discharging its contents, it is retained in position a few moments, in order that the atropin shall be retained. This injection should be done in the interval between the pains. He injects $\frac{1}{25}$ of a grain, dissolved in 10 or 15 drops of water. It caused no unpleasant symptoms in the eight cases reported; only a slight dryness of the throat, flushed face, and dimness of vision, which soon passed off. Probably the $\frac{1}{100}$ of a grain would be as effectual.

* No rational physician can believe that the 30th of Aloes or Podoph. will purge, or the 6th of Ipecac. vomit, nor can we expect the 30th of Belladonna to relax a sphincter.

used, and often the material substance itself. Give internally the 1^x or 3^x, and apply to the cervix with the finger a mixture of one grain of the solid extract, or ten drops of the tincture, in a drachm of glycerin, lard, or cosmoline, or a small ball of cotton can be saturated with the glycerole, and placed against the os.

Conium has some reputation in spasmodic conditions of the sphincters and circular-muscle-fibres. It has removed laryngismus and œsophageal spasm. I cannot, however, subscribe to the dogma that "any remedy will remove rigid os, if its other symptoms correspond," for if it *does* have all the other symptoms, it may not have that symptom of the os, and cannot therefore be completely homœopathic to the case.

Hyoscyamus and *Stramonium* both cause primarily paralysis of circular-muscle-fibres, and if the patient has other characteristic symptoms of these medicines, they may be indicated instead of Belladonna. I once removed a spasm of the cervix with *Solanum* (a close analogue of Belladonna), selecting it for the peculiar occipital headache and amaurotic symptoms.

Cactus causes, primarily, a spasm of the circular fibres of the heart, and has been successful in spasmodic dysmenorrhœa. If the woman has the cactus-heart-symptom, so well known, I believe it would remove a constriction of cervix. Perhaps it would be successful even if that symptom were not present. The 6th would be indicated, or even the high potencies.

Lachesis would be indicated, especially if its throat symptoms were present, and here the 30th, or the very highest potencies will be useful.

Caulophyllum irritates all circular muscles by its primary action; the constriction is *intermittently manifested*. This is precisely the condition of the cervix in the worst cases. The uterine pains will be very severe, the contractions of the fundus strong, but the cervix will also contract powerfully, and neutralize the uterine contractions, and the head cannot engage. In the 1^x or 2^x dilution (or trituration of Caulophyllin), it will promptly remove this condition. Larger and often-repeated doses would *aggravate;* they are only useful in general uterine atony.

Cimicifuga has a similar action, but the pains are irregular, the contractions irregular, and the cervical constriction irregular. One moment it would seem as if the os would sufficiently expand, the next it is firmly closed by spasm. Moreover, the patient is nervous, depressed in spirits, her hands and legs tremble, her pulse is quick and weak, and there is irregular twitching and choreic movements, and a dull heavy headache in the brow and eyes. When these symptoms occur, the attenuations from the 3^x to 6^x will be found useful; they should be repeated every few minutes.

Viburnum, from its wonderful power over spasmodic dysmenorrhœa, ought to prove very useful, and act very promptly in spasmodic stricture of the cervix, especially so when the patient has very violent pains, almost driving her distracted, attended by uncontrollable nervous excitement, cramps in the legs, thighs, and abdomen. It is primarily a uterine sedative; its spasmodic effects are secondary. Empirically it has acted best in appreciable doses. I have never observed it to remove spasmodic pains when given in the attenuations above the 2^x, but have seen the most happy effects from doses ranging from 10 to 30 drops of the tincture, or 1^x dilution, frequently repeated. No aggravation need be apprehended, for it is harmless, and I have known much larger doses used, with only pleasant results.

Morphia.—This medicine has been very successful in the hands of many physicians of all schools in certain cases of dystocia, but its sphere of action and the symptoms indicating its use have never been clearly defined. It has generally been prescribed for "excessive pain and uncontrollable nervousness," and in certain cases it has been remarked that a rapid and favorable termination of labor resulted from its administration. Some of my professional friends claim to get excellent effects from the 1^x trituration. In several cases I have been agreeably surprised at the favorable turn of affairs following similar doses. Instead of the *Sulphate* usually prescribed, I prefer the *Valerianate of morphia*, in the 1^x or 2^x trituration.

In the April number of the *Eclectic Medical Journal*, Prof. Edwin Freeman has a paper "On Morphia as a Partus Accelerator," and his views appear so clear and logical that I present them in full. He says:

"Eight or ten years ago I noticed in the *Medical and Surgical Reporter* of Philadelphia, reports of cases of rigidity of the os uteri favorably affected by the use of Morphine. A physician was called in the night to a case of labor. An examination revealed a rigid os. He concluded that the labor would be somewhat protracted, and that he might as well go home and enjoy a night's rest in his own bed. A moderate dose of Morphine was given to allay unnecessary pains, and he retired. In a very short time, however, he was aroused in haste and summoned to the bedside, which he barely reached in time to assist in birth. He was led to regard Morphine as a parturifacient, classing it with Ergot and other remedies of that kind. Subsequent observers of a similar action have regarded it as directed to the nerves supplying the lower parts of the cervix and os uteri, obtunding their sensitiveness, and thus preventing their reflex action upon the muscular tissues. This latter explanation probably approximates the truth, with the exceptions to be hereafter mentioned, so that the agent is rather a partus accelerator. Its action may also be referred to a lessening of the irritability of the muscular fibre, which irritability in that region is often increased by any unnatural condition of the part, such as previous cervical endometritis, granular os, wearing an instrument to support the uterus, a highly nervous and excitable temperament, too frequent examinations, etc. The muscular fibres of the uterus are of the involuntary kind, consisting of transverse fibres passing around the neck, acting as a kind of sphincter, and oblique and longitudinal fibres continuous with those of the middle layer of the muscular wall of the body of the organ. Involuntary muscular fibres are excited to activity by mechanical agencies; those of the alimentary canal by the contact of the contained alimentary material. They may be aroused to activity, or modified in their action, through the sympathetic system, or even through the cerebro-spinal system, but this is not the normal method. In like manner they are excited to unusual activity when the contained material is of an irritating nature, or when the canal itself is in a state of irritability, arterial blood being in excess, they are excited by the presence of ordinary material.

"In the progress of the evolution of the fœtus and the womb which contains it, the muscular fibres having become fully developed, the sphincter of the neck is gradually distended from above downwards by the gravitating pressure of the fœtus and amniotic bag that surrounds it, until at last only the external os remains to be loosed, and the uterus appears as a sac with an os and without a neck. At this latter stage of utero-gestation at the completion of the above process, and as its result, the child sinks down below its previous position low into the pelvis. The loosing of the transverse fibres still goes on, the settling of the child making space above, upon which the uterus contracts. The action thus set up in the body of the womb is continued to the oblique fibres of the neck, which, by then contracting, pull upon or stretch the constricting fibres of the os, still further enlarging that opening, until at last it is sufficiently open to allow the fœtal head to pass.

*   *   *   *   *   *   *

"Now if these sphincter fibres at the os be unduly excited, they may rigidly close it, or being less excited, they may seem relaxed, but as soon as the expulsive effort forces the child against the margin of the opening, contraction is set up, and it grasps it as rigidly as before. This is evidently the hyperexcited action of the non-striated muscular fibre, which ordinarily is excited to action by mechanical pressure.

"The objections to the theory that these contractions are produced entirely by direct reflex-motor action from the spinal cord, or even from the sympathetic, are: 1st. The fourth sacral nerve from the spinal cord joins the inferior hypogastric plexus of the sympathetic below where the uterine branches are given off, while above that point the spinal fibres join only the sacral sympathetic ganglia. 2d. The nerves distributed to the lower part of the womb are from that plexus, accompany the arteries, and belong to the sympathetic system, and their office is to regulate the functional activities of the womb by their influence over the circulation. 3d. There are no striated muscular fibres at the mouth of the womb to be affected by fibres of spinal nerves. Thus both these nerves and the sympathetic have, if any, only an indirect action. Now what is the probable method of action of Morphine? We give it or Opium in diarrhœa, and it lessens the peristaltic movement of the bowels, not through the nerves, but through the blood by direct action upon the fibres. So in the case of the uterus, its action is direct through the blood, lessening the irritability. The action of the fibres of the os are thus brought to harmonize with that of the rest of the womb, when it again begins, and labor proceeds without the previous interruption. I was called to a case of labor recently in which the os was soft to the touch, but quickly contracted as the pain came on. The patient remarked that she was a week in labor with her previous child. The 'waters' were coming away since the previous day. I waited awhile after having her get up and sit and walk a little, and then examined her, but there was no change, and the pains seemed lessening, although I gave her a dose of flu. Ext. of ergot. I gave her then Morphine sulph., gr. ¼ granule, and left her with directions to send for me as soon as the pains were severe. In an hour and a half I was called and found the labor progressing well, and she was speedily delivered of a child without further trouble."

But the *second stage of labor*, if extended beyond ten or twelve hours, presents symptoms which need attention. The pains become irregular, both in frequency and intensity, the fœtus seems to retrograde instead of advance, because the pains are not expulsive. "This local disorder is soon followed by violent trembling, the woman vomits bilious matter, she is uneasy and excited, and changes her position every moment, the skin is hot and dry, the pulse runs up to 100 or 150 per minute, the tongue is dry, and sordes on the teeth. The vagina and cervix are hot, sensitive to the touch, and a yellowish liquid escapes from them, which occasionally has a fetid odor, the pressure of the child's head on the neck of the bladder prevents the emission of urine." (Cazeaux.)

The above picture is one of an extreme case, a condition which I cannot imagine could obtain in the hands of a competent physician, for we have many remedies which can prevent such an array of symptoms, the chief of which are *Aconite, Baptisia, Cimicifuga, Caulophyllum, and Pulsatilla*. This will be seen by a brief comparison of the symptoms of these remedies with those of the abnormal condition. Besides these medicines I consider the *sitz-bath* to be of the greatest efficacy. Should I find a patient in the second stage of labor with the above serious symptoms, I would immediately give her the remedy indicated, and have her placed in a deep sitz-bath of a temperature of 100° F., and kept there till she felt faint, or till the natural pains and expulsive efforts came on actively. I would do this before resorting to the use of *forceps*, because the patient and the child would be better off for it.

But we will now consider the special causes that may bring about the above abnormal *second stage*, namely: *Slowness* or *feebleness of the contractions*.

This condition may occur at the very commencement of labor, and persist to the end, unless the appropriate remedy is given.

It is sometimes, but *not often*, due to *general* debility of the muscular system, for it is a well-known fact that the most feeble patients, as those in consumption, often have the most powerful pains, and the most rapid labor.

When the feeble pains are due to *general* atony, words of encouragement, small quantities of broth or beef-tea, or a few spoonfuls of wine, or Elixir of Coca, frequently repeated, are of great service.

Caulophyllum tincture in drop doses, or Caulophyllin, 1^x or $\frac{1}{8}$ grain pills, repeated every half hour, even before the *os* is dilated, will have an excellent effect in increasing the natural uterine action. This remedy seems to affect specifically all the muscular fibres of the womb in a manner simulating *natural* labor. This explains why it appears to *relax* the os in cases of rigidity.

Cimicifuga is best indicated where the pains, besides being feeble, are *irregular*, and are not always in the same place,—

now on one side, now on the other, now in front, then in the back. The *os* does not relax, and the woman is depressed in spirits, taciturn, and has jactitation of the limbs. The lower attenuations are indicated.

Pulsatilla is indicated for these changing, erratic pains, but the woman *weeps, scolds, frets,* and tosses about in bed, and is very "fidgety."

China is of great value in such cases, but it should be given in appreciable doses, namely, a teaspoonful of the *wine of Cinchona*, repeated every hour or two. Its alkaloid, *Quinine*, however, gives the best results. Many eminent obstetricians believe that Quinine possesses parturient power similar to Ergot, but I doubt its supposed excito-motor power over the uterus. It causes natural uterine pains, because it is secondarily homœopathic to muscular atony when due to inefficient blood-pressure, and decreased nervous supply. In many cases of general exhaustion, with cool sweating skin, feeble quick pulse, and very feeble *pains*, I have seen almost magical results from the administration of half a grain or a grain of Quinine every half hour. The pains begin immediately to increase, the pulse is stronger, and the woman becomes greatly encouraged. The obnoxious taste of the drug is readily disguised by mixing it with Syrup of glycyrrhiza (gr. 1 to 1 drachm), or the use of gelatin-coated pills, of a grain or a fraction of a grain each. In cases of alarming prostration, five or ten grains at a single dose has often seemed to rescue the woman from collapse.

I ought to add that the physician must not mistake the *condition* of the womb in cases of feeble pains. *Excessive distension of the uterine walls* often causes very feeble and irregular pains. The uterine walls become so thin that they are benumbed, and cannot contract while so distended. This condition may be diagnosed by an *examination*, when the bag of waters will be observed *not* to bulge during the pain. In such cases, if the *os is not rigid*, the remedy is to rupture the bag of waters, and *then*, if good pains do not set in, give one of the remedies just mentioned.

Slowness or *feebleness of the contractions* may depend upon sanguineous engorgement or plethora of the uterine tissues.

This condition may be diagnosed by the following signs: The pains are at first quite energetic, but soon diminish, both in frequency and intensity, the cervix uteri is soft, supple, and non-resistant, but the presenting part of the child does not engage during the pain, which latter is equally diffused *over the whole abdomen*. The phenomena of *general* plethora nearly always manifest themselves at the same time, the respiration becomes laborious, the pulse hard and full, and the pains irregular both in force and frequency. In these cases the old practice is, or was, to bleed, for it is very rarely resorted to now.

We have remedies, fortunately, that will control this condition without debilitating the system.

Gelsemium, if the face is flushed, the pulse *full*, but not *very hard*, the head feels heavy, the senses apathetic, and the pains feeble and irregular; it will soon remove the plethora if administered in the low dilutions.

Veratrum viride is indicated in a higher grade of congestive phenomena; when the head aches *violently*, the pulse is *very full, very hard and quick*, and convulsions or apoplexy threatens; and especially if the temperature of the patient is high—103° or 105°.

Prompt results will accrue from two or three drops of the tincture, or 1^x dil., repeated every hour, or oftener.

Cactus will be useful if the heart is violently excited, and there is a sense of *constriction* about it.

Bromide of soda (5 grains every half hour) has often removed this plethoric condition, when actual fever was *not* present, and there was great nervous erethism.

Aconite is not indicated, and should not be used unless the pulse is *small, hard* and *quick*, and there is great anxiety and fear of death.

If the feebleness of the contractions be solely due to uterine debility, a weakness of its muscular tissue, while the general muscular system is healthy and strong, we must select remedies which specifically *cause* this condition.

In such cases the woman will exert great strength; will "bear down" very powerfully, and pull violently on the hands of the attendants; but the physician will observe by "touching" that the head does not advance, although no obstacle to its progress is present.

Here we have admirable remedies in *Caulophyllum, Cimicifuga, Cannabis indica, Secale, Ustilago,* and *Phoradendron.* (See page 362.)

The same medicines are equally useful in *sudden cessation of the pains,* either from exhaustion, or some unknown cause. *Caulophyllum* is preferable where the pains are exceedingly feeble and far apart, or when they have been very violent, but intermittent, and then cease altogether. Give the tincture, ten to thirty drops every fifteen or twenty minutes, or the Caulophyllin in one-tenth or one-fourth grain doses as often, or until good pains appear.

Cimicifuga is superior to Caulophyllum, if the feeble and inefficient pains are irregular, both in intensity and recurrence. The general system partakes somewhat of the same aberration. It is a greater neurotic remedy than Caulophyllum, and its power over the uterus depends on its action through the spinal and sympathetic system. In cases of sudden suspension of pain and expulsive efforts due to depression of mind or nervous exhaustion, Cimicifuga is specific. The most efficient dose lies between the tincture and second dilution, in drop doses, or from the 1^x or 2^x trit. of macrotin, in one grain doses, repeated every fifteen or twenty minutes, until the pains are efficient.

Cannabis indica in doses of five or ten drops of the tincture repeated every twenty minutes has been found to increase inefficient pains or bring back suspended ones. It doubtless acts upon the enfeebled nerves which supply the uterus, stimulating them to normal action.

Electro-magnetism properly applied has been known to act well in such cases, after all medicines have failed. The *positive* pole should be applied to the lumbar region, the *negative* to the os, in the vagina, and a mild current (induced) should be applied between the pains, and suspended during their continuance, or, if no pains are present, the current should be applied for five minutes, with intervals of the same duration.

Secale has been used by the dominant school for nearly a century, for the purpose of inducing uterine expulsive efforts.

At first its powers were not understood and it was given in all stages of labor, when the pains were feeble or absent. It was found that in many cases its effects were unpleasant and dangerous, or no apparent effects were observed. As its action became better understood, it was found that if given before the os was dilated or easily dilatable, it caused violent, persistent, painful contractions, which were not only useless to expel the child, but were the means of its death.

The intense and unintermitting pressure cut off the fœtal circulation, and the child died of asphyxia *in utero*.

Even in cases where the os is dilated, the constant pressure caused by Ergot will produce the death of the child before it is thrust into the world. Its use among the best accoucheurs of the old school is now limited to those cases where, after the os is dilated, labor does not advance, owing to feeble expulsive power, the size of the child, or smallness of the passage. Many of the more cautious do not advise it until just before the head escapes from the vulva, and then only for the purpose of insuring a firm, final contraction, in order to prevent hæmorrhage, open or concealed, or the formation of an intrauterine clot. They prefer the use of *forceps*, for the purpose of facilitating the labor. In this I think they are judicious, for the safety of the child and the integrity of the perinæum are greatly enhanced by their skillful use.

If the pains are simply feeble and inefficient, the careful application of the pocket-forceps (page 370), in such a manner as to *imitate* the natural expulsive efforts, and *aid* the inefficient pains, is much more efficacious and desirable than a resort to such a powerful agent as Secale. If, however, the accoucheur has not the forceps with him (but he always should have), and the pains are very feeble, slow, and inefficient, or altogether absent, and the os is dilated, the presentation favorable, and perinæum distensible, Secale should be administered. I must here protest against the absurd advice given by some writers of our school, of giving the 30th or 200th in such cases. As well might we expect to *cause* diarrhœa with the 200th of Epsom salts. Only appreciable material doses of Ergot will excite or originate uterine contractions. Any observations to the contrary are based on a delusion or coincidence.

There are several methods of administering Ergot in such cases. (1) The old method of infusion—30 to 40 grains of the powder in an ounce or two of hot water; a teaspoonful every ten minutes, or all in three or four doses, half an hour apart; (2) 15 or 30 drops of a fluid extract at similar intervals; (3) a teaspoonful of the wine of Ergot, as above; or, (4) Ergotin in gelatin-coated pills, containing three grains each, one every quarter or half hour. (5) A hypodermic injection of 15 to 20 drops of Squibb's, or any other good *aqueous* extract, will act with greater certainty and promptness than any other method of administration. The drug should be suspended on the appearance of strong bearing-down pains.

Ustilago, or the Corn ergot—is supposed to have similar properties as Secale, but it has not been sufficiently tested to enable us to use it with confidence. The dose is the same as Secale.

Gossypium, or Cotton-root, is praised by many southern physicians as a good substitute for any of the above medicines. In several cases of slow labor in the last stage, I have given teaspoonful doses of the fluid extract. It appeared to stimulate to more efficient expulsive efforts. It is said that a decoction of the fresh root is the most efficient preparation, but the use of this would be restricted to the localities where the plant grows. We need further experience with Ustilago and Gossypium to determine their real powers.

Viscum album.*—This new remedy has lately been introduced as a partus accelerator. In my *Therapeutics of New Remedies* I refer to its use by Dr. Huber, of Germany. He used it successfully in *retained placenta, metrorrhagia,* and *menorrhagia,* and his

---

* Since the appearance of Dr. Long's statement, Dr. E. S. Crosier has investigated the botanical relations of the Mistletoe used by Dr. Long, and he announces in the *Louisville Medical News*, that it is not the Viscum album, but the *Phoradendron flavescens* of Nuttall. Therefore the Viscum album cannot be used for this purpose, although it may possess similar properties (see *New Remedies*, 4th edition), and resembles Cimicifuga in many respects.

Dr. A. G. Hobbs, contributes to the same journal a report of three cases, two of them obstetrical and one of menorrhagia, and says, "My experience with this parasite is that it acts more promptly and more decidedly as an oxytoxic than Ergot." (See also a case in *Amer. Jour. Obstets.*, July, 1857.)

cases appeared to show that the condition was that of uterine inertia. I have never tested it fully, and therefore cannot give my favorable testimony. But I will quote the testimony of Dr. Long, which seems conclusive.

"Dr. William H. Long, of Louisville, says that for ten years he has used the Mistletoe as an oxytoxic, having been led to do so from observing that farmers, in the part of the country where he formerly had practiced, were in the habit of giving Mistletoe to such of their domestic animals as failed to 'clean themselves,' or expel the placenta after the delivery of their young. In 1857 he first used an infusion in the case of labor, in which the second stage was delayed through inefficiency of the uterine action. Contractions followed in twenty minutes. He has since used it in decoction in a large number of cases, and does not recall an instance of its having failed to stimulate the uterus to contract.

"He believes in its superiority to Ergot—

"1st. Because it acts with more certainty and promptness.

"2d. That instead of producing a continuous or tonic contraction, as Ergot does, it stimulates the uterus to contractions that are natural, with regular intervals of rest. Consequently it can be used in any stage of labor, and in primiparæ where Ergot is not admissible.

"3d. It can always be procured fresh, does not deteriorate by keeping, and is easily prepared.

"He has used Viscum in many cases of menorrhagia and hæmorrhage from the uterus with gratifying results, and has taken pains in such cases to give Ergot and Mistletoe a competitive trial, with the object of testing their relative merits; he unhesitatingly pronounces in favor of the latter. Indeed, cases in which Ergot given in powder, decoction, and fluid extract failed to give any relief, the Viscum acted promptly.

"In post-partum hæmorrhage, the results have been no less satisfactory than in labor and menorrhagia, firm contractions of the uterus being secured in from twenty-five to fifty minutes after administering from one to two doses of the Mistletoe.

"According to Dr. Long, the remedy may be administered either as an infusion, tincture, or fluid extract, but he considers the latter to be the most convenient. The former he directs to be made by taking two ounces of the dried, or four ounces of the green leaves; pour on these one pint of boiling water, cover closely and allow to stand until cool enough to drink. Two or four ounces may be given at a dose, and repeated in twenty minutes if necessary. The green leaves impart a disagreeable taste that is lost in the process of drying.

"He has also used an alcoholic tincture made by taking eight ounces of the dried leaves and saturating them with boiling water, and adding alcohol to make one pint; but he does not think this as efficient as either the decoction or fluid extract. It should stand ten days before ready for use. Viscum makes a fluid extract of a dark-brown color, which possesses all the virtues of the parasite.

"The best time for gathering the Mistletoe is in November, after a few frosts have fallen, and before the sap freezes, though it may be gathered and used at any period of the year. When gathered, it should be at once spread out to dry, as it will mould in a very short time if kept in a box or sack. It is best to dry it in the shade.

"Viscum abounds in the Western country, and is found in greatest quantities on the walnut and elm trees, though it grows sparingly on a few others, as the red and black locust, oak, etc. So far as Dr. Long is aware, there is no difference in its properties or strength made by the kind of tree on which it grows."

Sudden cessation of the pains may be caused by influences from without. Disagreeable impressions occurring during labor— like the presence of disagreeable persons, the arrival of a physician, especially if he be not the one expected—may determine the cessation of the pains.

Aconite is the remedy if the cause is akin to *fright*.

Chamomilla, if the patient is vexed or made angry.

Other remedies may be selected according to the causative influence, not because the medicine affects the uterus directly, but because it acts upon the mental or moral origin.

During the progress of labor *acute pains in some other portions of the body* may occur and arrest the labor until the extraneous pain is removed.

Violent vomiting may occur, and will require the administration of Ipecac., Iris, Nux vom., Stibium or Verat. alb.

Lumbar pains, severe and sharp, may require Belladonna, Cimicifuga, or Rhus.

Colic may require Chamomilla, Dioscorea or Colocynth.

Violent cramps in the thighs, legs or calves, so graphically described by Prof. Meigs, may set in with such severity as to totally arrest the uterine contractions. They are supposed to be caused by the pressure of the child's head upon the sacral nerves, and no remedy is capable of relaxing them until the pressure is removed. Here the forceps should be promptly used, and Chloroform or Ether given. Should the cramps continue—as they sometimes do after the pressure is removed— Arnica, Colocynth, Verat. alb., or Viburnum are useful.

Congestion to the head has been known to occur suddenly during labor, with sudden disappearance of the pains. In such cases, nurses say, "The pains have gone to the head." I have met with several such cases. The patient suddenly puts her hands to her head—the pain is severe; there is flushed face, vertigo, dimness of vision, throbbing, ringing in the ears, pulse full, hard, etc.

One such case was soon relieved by ten drops of tincture Verat. viride; and another by the administration of thirty grains of Bromide of lithia, which effected amelioration in thirty minutes.

In the former case Verat. viride was selected because the pulse was *very hard*. In the latter the pulse was *not* hard, but *full* and *soft*.

Solanum, Gelsemium, and *Belladonna* will be useful if other symptoms call for them.

Violent pains in the uterus, or pelvis, from pressure of the child's head, may be so intense that the woman cannot make any effort, at a time when her efforts are necessary to the termination of labor.

I have met with cases where the pains were so spasmodic, cramp-like, and agonizing, that labor actually came to a standstill. In some cases a hot sitz-bath will give almost immediate relief, so much that the woman will beg to be allowed to remain in the water. In other cases there is nothing to be done but to remove the child with *forceps,* or give Chloroform to induce anæsthesia. In a few cases I have known medicine to allay these pains. In one instance a teaspoonful of tincture of *Caulophyllum* gave relief in ten minutes, probably by changing the position of the child's head. In another case *Viburnum opulus,* in a similar dose, relieved the pain after three doses half an hour apart. Theoretically *Secale* 30th ought to relieve the *agonizing, constant* pain which sometimes occurs.

Cases are on record where Hyoscyamus, Coffea, and Chamomilla are said to have relieved these unnatural pains.

Irregularity of the pains.—Under this head Cazeaux describes certain kinds of pains, which he also terms *uterine tetanic* pains.

He mentions several varieties:

1. "There is not a complete and perfect interval between them, they are continuous, and only interrupted by the paroxysms, during which the intensity of the suffering is terrible." This variety, as above stated, would call for Secale[30], according to a strict application of the law of *similia*. Here, evidently,

Chloroform ought to be given, until the pains cease or give way to natural ones. If Chloroform was not admissible I would try *Amyl, Calabar,* or *Chloral hydrate,* either of which, in proper doses, should, in a short time, ameliorate the sufferings.

2. "The pains return, it is true, at intervals, but sometimes it is only the fundus, again one of the cornua, and at others, some part of the uterine body, which contracts spasmodically, while the remainder scarcely does so at all."

They have no effect on the progress of labor, except to retard it. The hand on the abdomen notes the irregular contractions of various parts of the uterus. The membranes do not bulge, nor the head press down. In these cases the warm *hip-bath* may do much good.

Cimicifuga may prove an excellent remedy if the woman is nervous and depressed, the whole body jerks, and the limbs twitch (choreic motions). (See Cannabis indica.)

Gelsemium has proved the curative remedy in several such cases under my care; and recently Dr. Fauntleroy reported a case to the Virginia Medical Society, which verifies my experience. He writes: "I was called in consultation, and the following history was elicited: In the *three* previous confinements, from the exhaustive continuance, for two or three days, of the inefficient contractions, marked by frequent pulse, coated tongue, and mental wandering, the doctor had been forced to relieve his patient by a resort to instruments. When called upon, the labor had commenced, the os uteri was partially dilated, *not at all rigid,* but the contractions evidently involved different planes of the uterine muscular tissue, first in one part, then in another. The writer suggested the use of Gelsemium. Eight drops of the fluid extract were given every hour. After the second dose the uterine contractions became more general, and when the patient had taken eight doses, she was delivered by the unaided forces of nature, of a large healthy child."

Pulsatilla.—If this medicine is ever indicated in unnatural labor, it is applicable in this variety. The genius of this remedy makes it applicable, not only for irregular contractions of the uterine muscles, but contractions which *fly* from one set of muscles to another in the uterus, or from the uterine muscles to those of other parts of the body. The mental state of the patient affords valuable indications for Pulsatilla. With irregular, inefficient pains the woman often weeps, complains, is greatly agitated, and cannot be encouraged. Here Pulsatilla will be indicated.

Aconite will be useful if the patient has an uncontrollable anxiety and fear of death, the skin is cold, and the pulse small and feeble (primary), or she becomes hot, feverish, with hard small pulse, thirsty, delirious, and has slight convulsive motions. (Secondary.)

Ignatia covers many of the symptoms of this condition.

Dioscorea may prove useful, especially if the uterine muscular-cramps suddenly cease, and appear in the bowels, hands, or feet. (See *Symptomatology of New Remedies*.) Before closing this chapter I cannot omit to mention a remedy, which under certain circumstances proves most efficient and harmless in the treatment of *false labor*, or irregular uterine contraction.

I allude to *Opium and its preparations*. They are not homœopathic to this condition, any more than Chloroform or Ether, but act as palliatives, giving the tortured organ needful rest, after which it takes up its normal action.

Cazeaux says: "Under the influence of Opium, given in the form of enema (into the rectum), twenty or thirty drops in a few ounces of water, the pains entirely disappear in the course of half an hour. During this period the patient generally slumbers, and then the good pains, that is, the natural and regular ones, come on, and the labor terminates happily."

In several cases of peculiar obstinacy, where neither the homœopathic remedy nor Chloroform changed the unnatural character of the pains, I resorted to an enema of twenty drops of McMunn's Elixir of Opium, and the improvement set in in less than half an hour, and after an hour or two of sleep, natural labor commenced in earnest. In one case, two doses, half an hour apart, of one-eighth of a grain of Morphiæ acetas had the same effect. (See Dr. Freeman's observations on Morphia, p. 274.)

Position.—A woman in labor should not be obliged to lie in one position, unless it is appropriate, and she desires to. To confine a woman to the left-side position in all cases, would be to greatly retard labor in many cases. In *lateral oblique* positions of the uterus, or oblique presentations of the head, the woman should lie to the side opposite to that of the fundus uteri. This allows the uterus to fall on that side, and changes the abnormal presentation of the head to a normal one.

In *anterior obliquity*, the proper position is on the back, with the head as low as comfortable. This obliquity may also be removed by a properly abjusted bandage worn during the last month of pregnancy, and also during labor. It should be so applied as to hold the womb upward, and press it backward. A finger hooked into the os, and drawing the anterior lip forward will often remedy this condition.

In *posterior obliquity* (see Cazeaux, page 714), the woman ought to remain seated or standing, or, if possible, reclining a little forward. I have in two cases of this nature delivered the woman while she was kneeling by the side of a bed or chair, with the body bent forward. The hands of the accoucheur, reaching from behind, and pressing on the child's head just above the pubes, assists in aiding the head to engage.

In all these obliquities the head is often greatly obstructed by the rim of the cervix getting between the head and the passages. This should be pushed forcibly back, *between* the pains, and held there, if possible, *during* the pains, until it will recede permanently.

Abdominal pressure, made by the hands of the physician or nurse, or by means of a properly-adjusted bandage, will often facilitate a slow and tedious labor. The external pressure should be made during the pain, and if possible, the pressure should be maintained in the intervals to keep what was gained by each pain. Many obstetricians of the present day strongly advocate this method. Unfortunately, some women cannot bear any pressure on the abdomen during labor, unless they are under the influence of Chloroform, Ether; or Chloral hydrate.

The Forceps.—I do not propose to mention those Forceps whose sole use in difficult or abnormal labor is to *extract the child by force*, but shall briefly mention a variety which can be used as a simple and valuable aid to the natural or deficient pains.

My first experience with forceps was with Davis's, afterwards with Simpson's, but recently I have used Comstock's with decided satisfaction, especially when the head was high in the pelvis. When the head has passed the brim and the obstruction is below the promontory of the sacrum, unless there is some very great obstacle to overcome, I prefer Roler's.

But from the first I was struck with the absence of any instrument, which, without being a *compressor*, or powerful *tractor*, would be an *aid* to the mother in the expulsion of the head from the lower strait, or when it rests on the perinæum.

Every practitioner has had cases where the head comes naturally down upon the perinæum, and then from some cause, as fatigue or exhaustion of the mother, a head that would not readily mould, rigidity of the soft parts or perinæum, or dryness of the passage, refuse to progress with that celerity which was desirable.

I hold with Dr. Goodell, of Philadelphia, that it is only very rarely that there is any necessity for waiting hour after hour in slow labors, when we can facilitate it by means safe to the mother and child.

But must we always use the heavy and formidable instruments which up to this time have been invented? The modern tendency of all the instruments used by surgeons, has been toward lightness and delicacy; not only in surgery but in all the arts, and in every department of labor where machinery is used. Why has it not been so in midwifery?

It is only within a few years that Roler's light forceps were introduced, and this has met with much opposition. Years before I saw Roler's I had made a drawing of a small light forceps, which should be of the same shape as Davis's only shorter and very much lighter; but I shrank from intruding my idea upon the profession. However, when I saw the extremely light and delicate forceps, having only a "scissors handle," invented by Dr. Newman, of Denver, I hunted up my old drawing, in which the handle was the ordinary club-shaped one of Davis's and Roler's latest. I then conceived the idea of modelling the handle of my small forceps so that it would resemble that found upon some of the older styles of pistols. I finally adopted the following shape, as shown by the accompanying cut.

Fig. 54.

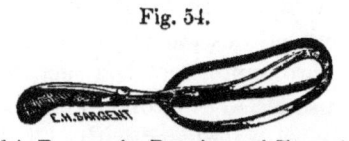

Hale's Forceps for Rotation and Extension.

I have named my instrument the "Pocket Forceps," and a brief description is as follows:

Length of blades from point of crossing, six inches. Total length, ten inches. Length of handles, three and one-half inches. A gradual pelvic curve of two inches, beginning near the extremity of the shanks. Breadth of the cephalic portion averages one and one-half inches. The distance across the widest part when the instrument is closed is two and three-quarter inches, the points being separated one-half of an inch. The branches fit easily in a "button lock." The diameter of the closed handles is one inch. The fenestræ are nearly elliptical; their length the same as most other forceps. Total weight only five and one-half ounces. These forceps are all nickel-plated.*

[By way of comparison I give the relative dimensions of my forceps, Comstock's, and Elliot's. The total length of Comstock's is *fourteen inches;* weight, *thirteen ounces.* Total length of Elliot's, *seventeen* inches; weight, *twenty* ounces.]

---

* Since the introduction of my small forceps, many physicians of both schools have urged me to enlarge them so as to give them more *traction-power*. They highly approve of the shape of the handle as being the most convenient ever introduced. I have complied with their solicitations, and have designed a forceps which I have named *Hale's Larger Traction Forceps*, of which the following is a description:

Length of blades, *six* inches; total length, *fifteen* inches. Length of handle, *four and a half* inches. Length of curve, and breadth of cephalic portion and other measurements same as small forceps. The handle is placed more at an angle with the blades than in the small forceps, and its diameter is *one inch* near the lock, and one inch and a half at the butt. The shank between the blades and the lock diverges as in Comstock's. The weight about *ten* ounces. This forceps can be substituted for any one of the large forceps, as it possesses traction-power equal to any one of them, and is much more convenient on account of its peculiar handle. (Messrs. Boericke & Tafel will supply either instrument to order.)

It will be seen that while I have followed the general outline of Roler's blades, I have *increased their curve at the point of divergence from the lock*, giving an opportunity for the placing of the forefinger as a guide and slight tractor. The shape of the handle I claim to be altogether original. It affords the best grasp for the hand, and does not interfere with its free movement.

Its small size enables the physician to carry it in his pocket. Its appearance is not in the least formidable. It can readily be applied in nearly all the various positions in which women place themselves during labor, and thus afford, with but the slightest show of an operation, very important assistance during the last stage.

It can be used to rectify those annoying *oblique* positions of the head, which, even when the head is resting on the perinæum, retards its progress. Here it is used as a *lateral lever*, as recommended by Meigs. In cases where the head is placed *transversely* to the lower pelvic strait it can be used successfully to *rotate* the occiput under the pubic arch. In face presentations, these forceps can be used effectually to cause the chin to escape more readily from under the pubes, or assist it in escaping from the perinæum.

In *occipito-posterior* positions this little instrument will greatly assist in causing the *chin to approach the breast*, and allow the vertex to escape more readily. Even in that most natural phase of labor, the *occipito-anterior*, the attendant and patient are annoyed and fatigued by the *tendency of the head to retrograde*, or "slip back," at each pain, just at the time when it seems as if it was about to escape from the vulva, or reach a position where it could be grasped by the hand. How often have we waited hours for the escape of the head under these circumstances. We feel that all that is wanting is a *little harder pain*. We know that if we only had one hand on the head we could extract it, but just as we seem able to grasp it it slides back into the pelvis. It is in just such cases that these small forceps are a great assistance to both physician and patient.

I here wish it to be distinctly understood that I do not recommend my forceps as an instrument for *forcibly extracting the child*, but for guiding the head, for effecting *rotation, flexion* and *extension*, and *assisting weak or even normal pains*.

For the purpose of forcibly impressing upon physicians the value of my short forceps, I quote from a recent lecture by Dr. E. W. Sawyer, of Chicago, who claims for it great value in *saving the perinaum* by effecting *flexion* at a certain period of labor. He says:

"The remarks which I have to offer, relate to those labors in which the vertex is in advance, and only to that stage of labor when the head has almost passed through the bony portion of the obstetric canal, but is still opposed chiefly by the woman's soft parts at the floor of the pelvis, and the outlet of the canal.

"It will perhaps elucidate the point I hope to establish, if I may be permitted to describe the manner in which the forces of the woman will complete the expulsion of the head, if no interference is offered by the attendant.

"At the moment we speak of, the antero-posterior diameter of the fœtal head approximately corresponds to the same diameter of the parturient canal. In the great preponderance of labors, the back of the head looks upward—the woman being upon the back. A little further advance brings the vertex to look through the vulvar aperture, and the nape of the neck in contact with the inner surface of the pubes; while that part of the occiput just inferior to the protuberance, is lodged against the sides of the pubic arch. The occiput is too broad to be received into the pubic arch as far as its summit, as one may convince himself by sweeping the tip of the finger between the inferior border of the symphysis and the head, during the expulsion of the latter.

"Till this time the fœtal head has been in a state of flexion; but when the occipital plane becomes arrested against the pubic arch, the frontal part of the head receives the propelling force more directly, and is soon in advance. The head is now made to describe a movement of extension, or evolution, by which it becomes unfolded into the world, around the point of the pubic arch, against which the occiput is lodged, as around a pivot.

"Two forces operate to produce this extension. The propelling power of the uterus and its auxiliaries, the *vis-a-tergo*, advance the head until the larger portion of the forehead looks over the margin of the perinæum. When the soft parts become still further distended by the fœtal head, the perinæum draws itself backward over the face, urging forward successively the margin of the orbit, the malar prominences, nose, lips, and chin, the part of the head to be freed last being that part which was lodged against the pubic arch. This retraction of the perinæum is the second force which extends the head.

"Such is an outline of the manner in which the head is delivered by nature. A movement which, while it succeeds in its object, at the same time jeopardizes the woman's soft parts. This is even more apparent when we recall the successively increasing dimensions of the head which pass through the vulvar outlet. I have given the name pubo-facial diameter to that axis, one end of which rests upon the highest part of the pubic arch, while the opposite extremity is lost upon different parts of the face. Their names indicate the limits more exactly. The length of these several axes show the extent to which the vulvar aperture is opened to give exit to the head. Thus the pubo-frontal, 4½ inches; pubo-malar, or nasal, 4¾ inches; pubo-mental, 5¼ to 5½ inches. I believe it practically impossible for the vulva of the primiparous woman to be stretched to this degree without a rupture of the perinæum occurring. I am aware that a primipara may be delivered of a large child, and her perinæum be left intact, but this is because the judicious interference of her attendant compelled the head to pass out in a shorter axis than nature can do if left to herself.

"I have had an opportunity of an occular demonstration of the moment and manner in which the perinæum was torn. Just as the upper margin of the orbits looked over the edge of the perinæum, the little fold of mucous membrane known as the frouchette, or frænum, gave way; this tear was continuously deepened by the malar prominences and chin. This is, I think, the usual order. Writers have described, in exceptional cases, its first giving way at the centre, but all agree that it tears on the median line.

"I know some hold that it is the bis-acromial diameter, the shoulders, which causes the tear. But I cannot believe that the perinæum, left absolutely intact by the head, will be torn first by the shoulders, under the care which the woman always receives. Such an accident can always be averted by delivering the pubic shoulder first.

"The interference which I would recommend to anticipate the hazardous stretching of the vulva is, in a word, of a nature to hold the head in a state of extreme flexion, and force it to pass through the vulva in a diameter a little superior to the pubo-frontal, and which has a length in the full-grown child of about four inches.

"I cannot assume that preventing the extension of the head at this time is an original procedure; only that its significance is not generally understood. Many practitioners interfere during this stage of labor, and really prevent the complete extension of the head. Thus some hook the index finger over the chin through the woman's rectum. It would seem at first sight that they attempted just what we would prevent, that is, extension of the head; but really they accomplish what we have recommended, for the thumb of the same hand is pressed upon the perinæum, which is being bulged out by the forehead. In this way, with the face held between the thumb and finger, in easy cases, the sinciput is kept back.

"Others apply the hand to the perinæum in such a way that the commissure between the thumb and finger corresponds with the posterior commissure of the vulva. Judicious support continuously supplied is of the greatest advantage, but the object to be attained by this palmar pressure is not always understood, for I have often been told to hold the forehead back for a time to allow the perinæum to become thin and the vulva more easily stretched. Possibly this is accomplished, but the greatest advantage comes from the flexion into which the head has, by this means, been forced.

"In addition to holding the sinciput back with one hand, others attempt to tease the occiput forward with the fingers of the other hand applied to the head just in front of the symphysis pubis.

"Besides the objections to the introduction of the finger in the rectum, it can be said against all these measures that they are not sufficient, in the majority of labors, to hold the head in a state of flexion.

"I am assured that this can be most certainly and easily accomplished with the forceps. For a long time I have been using my short forceps for this purpose with the most satisfactory results."

I will add that Dr. Sawyer's forceps are similar to my short forceps, with the exception of the handle, which, like Comstock's, has a hook to each half, which projects outward. I consider the "pistol handle," as applied to obstetric forceps, as one of the greatest advances in the improvement of that instrument. When placed upon the large forceps, it will give to them a strength, a convenience, and a facility for manipulation which will render it universally popular.

Dr. Sawyer gives correctly the proper method of the use of the small forceps:

"Let us assume that the forceps is to be used to flex the head, and to hold it in a state of flexion. At this stage in the labor the woman is usually upon the back. Her position need not be changed, only to have the limbs strongly flexed in the lithotomy position. Nor is it necessary for the operator's body to be in a line with the pelvic axis. The introduction and locking of the blades is a matter of extreme simplicity. While the head is loosely held, the handles should be elevated so as to nearly approach the symphysis. Choose the interval of uterine action. Now clasp the head firmly in the blades, and slowly depress the handles until the edge of the perinæum is approached. The following diagram illustrates this movement of forced flexion:

Fig. 55.

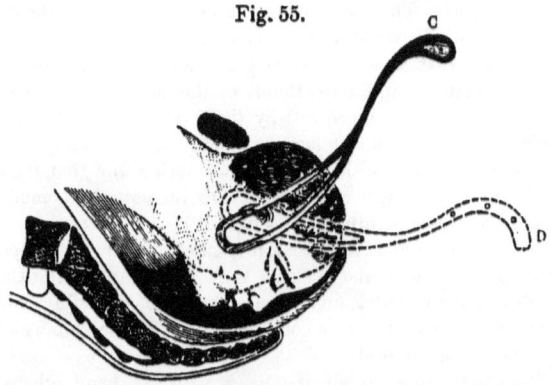

"C shows the forceps applied. The handle is to be depressed to D, when the chin will recede to the dotted line.

"It may be that one movement is not sufficient to flex the head; this is the more likely to result if the operator is overtaken by a contraction of the uterus. It is only necessary, in this event, to repeat the elevation and depression of the handles, and the operator will have the satisfaction of feeling the sinciput recede, when a moment before it was causing the perinæum to bulge out in a most threatening manner.

"The most important step in the operation remains to be mentioned. When the head has been flexed, the hold of the instrument should be relaxed, and the handles elevated to that degree that the general axis of the handle is nearly perpendicular to the plane of the bed. It is in this position that the head is to be held if the operator waits for the uterus to complete the delivery; and it is in this direction only that the operator lifts the head if he sees fit to make traction.

"If, in addition to this use of the forceps, the ends of the fingers of the disengaged hand are placed upon the head in such a way that the convex surfaces of the nails rest upon the edge of the perinæum, and during the interval of uterine action gentle efforts are made to tense back this edge, and to prevent it from being caught upon the advancing head, the operator has by his conjoined manipulation given the soft parts the greatest possible security.

"Finally, in an exceptional case of posterior position of the occiput, which refused to rotate forward to appear beneath the pubic arch, I was able by reversing the movement I have described, to lift the head into marked flexion, and to deliver a primipara of a large child without injury to the soft parts.

"As injuries of the soft parts, I have in the foregoing remarks alluded particularly to those rents which are apparent upon an inspection of the perinæum. But it seems reasonable that the liability to those not infrequent tears of the mucous membrane about the outlet would be lessened by the same measures which would prevent the dangerous distension of the perinæum."

INDEX.

Absence of uterus, 73.
  ovaries, 69
  vagina, 206
Atrophy of ovaries, 70
  of uterus, 77
Atresia of uterus, 74
  cervix, 74
  vagina, 204
Amenorrhœa, 175
Anteversion of the uterus, 110
Anteflexion of the uterus, 110
Anus, fissure of, 232
Amputation of cervix, 144
Applicator, Sims's cervical, 168
  Hale's intrauterine, 165
Abrasion of os uteri, 169
Areolar hyperplasia of uterus, 177
Abnormal shapes of cervix, 194
  os uteri, 194
Abortion a cause of sterility, 261
Albuminuria, 317
  treatment of, 319

Ball's operation for strictures, 106
Beebe's pessary, 135, 137, 138
Bougies, slippery elm, 160
  medicated gelatin, 162
Bozeman's dressing-forceps, 163
Bladder, irritable, 223

Causes of sterility, 33
Constitutional causes of sterility, 60
Change of climate, 63
Cervix uteri, stricture of, 79
  occlusion of, 75
  stenosis, 79
  division of, 80
  Peaslee's operation on, 102
  dilatation of, 107
  Hunter's dilator, 107
  Ball's rapid dilatation of, 106
  rapid and forcible, 108
  amputation, 144
  elongation, 143
  œdema, 151
  lacerations of, 201
  enlargement, 177
  abnormal shapes of, 193
  conoidal, 197
  "smashed hat," 195
  rigidity of, in labor, 267
Climate, influence of, 63
Convulsions, puerperal, 248
Chambers's intrauterine stem pessaries, 111
Cotton pessaries, 142, 145
Cloth tents, 159
Cervical protector (Wylie's), 157
  applicator, Sims's, 168
  Hale's, 168
  hyperplasia, 177
Curette, use of, 163
Calendula glycerole, 170
Conical cervix, 197
Crescentic os, 194

Coition, improper time for, 253
  position during, 256
  during disease, 254
  conduct after, 258
Copulation, 257
Clitoris, diseases of, 216
  enlargement of, 216
  abnormal irritability, 218
Carbonic acid gas douche, 276
Cases of retroversion during pregnancy, 305
Cases of sterility 199, 219
  fissure of anus, 231, 232
Cystitis, 223
Cystic degeneration, 163
Catarrh, vesicle, 223
Constipation during pregnancy, 249
Conception, 235
  prevention of, 259

Displacements of uterus, 122
Diet, improper, as a cause, 64
  during pregnancy, in obesity, 61
Degeneration of ovaries, 70
Dropsy of ovaries, 71
  during pregnancy, 244
Dysmenorrhœa, 80, 174
  pseudo-membranes, 175
Dilatation, rapid and forcible of cervix, 107
Dressings, medicated, for ulcerations, etc., 171
Depletion, local, curing sterility, 186
Diabetes, 222
Dress, improprieties in, 248
DISORDERS OF PREGNANCY, 283
DYSTOCIA, 283
  preface, 281
  due to fœtus, 255
  fruit diet for, 238
Discission of cervix, 90

Erotism, 68
Exercise, want of, 64
Elongation of cervix, 144
Elevators for retroversion, 141
Endometritis, 154
  fundal, 154
Erosion of cervix and os, 169
Enlargement of uterus and cervix, 177
Engorgement, chronic of, 177
Enemas of hot water, 183
Eruptions on vulva, 211
Electricity in areolar hyperplasia, 190
  in sterility, 275

Frigidity, 68
Fallopian tubes, stricture of, 72
Fundal endometritis, 154
Fitch's sound, 75
Flexions of uterus, 110
  cervix, 110
  diagnosis of, 115
  rapid and forcible dilatation of, 108
Fibroid tumors, 149
  ergot in, 152

INDEX.

Fibroid, diet in, 153
Forceps. Bozeman's dressing, 163
    Hale's pocket,
Fissure of vulva or vagina, 210
    urethra. 229
    anus, 231
    of rectum. 232
Fistula, vesico-vaginal. 229
Fœtus in utero, medication of, 334
Fruit diet in dystocia, 340

Galvanic pessary, 112, 122
Granular os and cervix, 169
Galvanism in sterility, 275
Gymnastics, 276

Hæmatometra, 75
Hale's intrauterine applicator, 168
    speculum. 105, 212
    forceps, 370
Hysterotomies, 83
    White's 104
    Skeene's, 195
    Peaslee's 101
    Simpson's, 83
Hypertrophy of uterus, 177
    cervix, 177
Hot water enemas in metritis, 183
Hymen, imperforate, 208
Hysterical diathesis, 209, 288
Hæmorrhoids, 231
Hygienic causes of sterility, 248
Hydropathy, 276
Hydrotherapy in dystocia, 299
Hardness of the bones of fœtus, 339
Hunter's flexion straightener, 107

Illustrations, list of, 11
Inordinate sexual intercourse, 63, 255
Improper diet, exercise, etc., 64
Incompatibility, 67
Inflammation of ovaries, 70
    Fallopian tubes, 73
    uterus, chronic, 154
    vagina, 215
    bladder, 223
    urethra, 227
Inversion of uterus, 147
Irregularities, menstrual, 176
Irritable tubercle of vagina, 210
Imperfect development of ovaries, 69
    uterus, 76
Imperforate hymen, 208
Intrauterine pessary, 111, 112
    stem, 111, 120
Insemination, 258
    abortive, 255
Intrauterine tampons, 171, 174
    packing, 160
    syringes, 162
    bougies, 162
    flexible medicated pencils, 165
    suppositories, 162
    applicator, 168
Iodine in areolar hyperplasia, 188
Incision of cervix uteri, 82, 98
Internal os, calibre of, 98

Kumyss, nutritive value of, 192

Lacerations of cervix, 201
Lateroflexion of uterus, 142
Leucorrhœa, 174
    vaginal, 214

Leeches in engorgement of the uterus, 183
    curing sterility, 186
Local depletion in chronic metritis, 183
    curing sterility, 186
Labor, slowness of, 348
    Phoradendron in, 362
    cessation of pains during, 364
    vomiting during, 364
    cramps in legs during, 364
    congestion to head during, 364
    irregularity of pains during, 365
    position during, 367
    obliquity of uterus during, 368
    forceps in difficult, 368
    pains, cessation of, 364
    irregular. 365
    violent, 365

Medicinal cause of sterility, 55
Medicines useful in sterility, 263
Mercurialization, 61
Mineral waters, 64, 123
Menses, retention of, 75
    suppression of, 175
    irregularities, 176
    frequent and profuse, 176
    scanty, 176
    delaying, 176
Metrorrhagia, 176
Menstruation, imprudence during, 251
Metritis, chronic, 176
Mammary irritation, 277
Medication of fœtus in utero, 334
Medicated pencils, 165

Neuromata of vagina, 209, 211
Neuralgia of vagina, 209
Non-retaining vagina, 206
Nutritive treatment, 190

Ovulation (introductory), 19
Obesity, 61
Ovaries, absence of, 69
    imperfect development, 69
    atrophy, 70
    inflammation, 70
    degeneration, 70
    tumors, 70
    dropsy of, 71
Œdema of cervix, 181
Os uteri, abnormal shapes of, 194
    external, stenosis, 99
    "pinhole," 194, 198
    crescentic, 194
    closed, 194
    internal, stenosis of, 98
Ovarian remedies for sterility, 263

Predisponent causes of sterility, 59
Plethora. 61
Prostitution, 62
Psychical causes, 67
Prolapsus uteri, 143
Procedentia uteri, 144
Packing the uterus, 160
Phoradendron in dystocia, 362
Pessaries, intrauterine, 74, 120
    galvanic, 112, 122
    Chamber's stem, 111
    Thomas's anteversion, 113
    retroversion, 119
    Albert Smith's, 119
    Jackson's, 119
    Beebe's, 135
    cotton ball, 142

378 INDEX.

Pessaries for lateroflexion, 143
    elastic ring, 144
    Zwang's pessary, 145
    Studley's, 119
    in conoidal cervix, 197
    concavo-convex, 195
Polypi of the uterus, 149
Polymnia uvedalia, 157
Porte-tampon, 174
Pinhole os, 194, 195
Peaslee's uterotome, 196
    operation for stenosis, 101
Pencils, medicated elastic, 165
PREGNANCY, DISORDERS OF, 283
    vomiting of, 302
    retroversion during, 305
PAINFUL AND DIFFICULT LABOR, 251
Preface to Sterility, 3
Preface to Dystocia, 251

Retention of menses, 75
Retroversion and flexion, 110
    a cause of vomiting in pregnancy, 304
Renal causes of sterility, 222
Rectal causes of sterility, 231
Rectum, ulceration of, 232
    fissure of, 232
    speculum, 212
Remedies for sterility, 263
    for dystocia, 256

Scoop, Thomas's, 163
Statistics of sterility, 56
Scrofula, 60
Syphilis, 60
Sexual intercourse, inordinate, 63
    in uterine hyperplasia, 153
Stricture of Fallopian tubes, 72
    uterine cavity, 75
    cervix uteri, 79
    vagina, 205
Subinvolution of uterus, 177, 182
Sound, Fitch's measuring, 75
    Jennison's exploring, 115
    Sim's and Simpson's, 141
Simpson's hysterotome, 91
Skeene's hysterotome, 193
Sims's operation for stricture of cervix uteri, 79
Speculum, Hale's 105, 212, 242
Stenosis of cervix, 79
    internal os, 95
Slippery-elm tents, 112
    bougies, 160
Syringes, intrauterine, 156
Suppositories, medicated, 162
    in vaginismus, 210
Scarificator, Butler's spear, 154
Semen, nature of, 235
    injection into uterus, 275
Spermatozoa, 236
    examination for, 240
    passage to ovule, 243
Special disorders of pregnancy, 302

Tent, cloth, 159
Tents, complete, 161
    slippery-elm, 169
    tupelo, 196
Twin-births, 62

Tumors of ovaries, 71
    of uterus, 149
    of meatus, 229
Tampons, cotton, intrauterine, 171, 173
    in vaginitis, 215
    in leucorrhœa, 216
Tenaculum, Nelson's, 105
Thermometer, clinical, 196
Temperature forbidding operations, 196
Tortuous, cervical canal, 195
Therapeutics of sterility, 262
Trachelotomy, 82

Uterus, absence of, 73
    atrophy, 77
    imperfect development, 76
    atresia, 74
    displacement of, 110
    anteversion, 110
    flexions, 110
    retroversion and flexion, 114
    during pregnancy, 305
    lateroflexion, 142
    prolapsus, 143
    elevation, 146
    inversion, 147
    tumors of, 149
    polypi, 149
    areolar hyperplasia, 177
    hypertrophy, enlargement, etc., 177
    subinvolution of, 182
    erosions, 169
    ulceration, 169
    abrasions, 169
Uterine elevator, Sims's and Gardner's, 141
    sounds, Sims's and Simpson's, 141
    applicator, 168
    pessaries, 111, 144
    medicines for sterility, 263
Ulcers of vagina, 210
    uterus and cervix, 169
Uterotome, Peaslee's, 196
Uræmia, 325
Urethra, irritable, 227
    dilatation of, 227
    caruncles of, 228
    granular erosion of the mucous membrane, 229
    fissure of the, 229
    vesico-vaginal fistula of, 229
    vascular tumor of the meatus, 229
Urethral catarrh, 227

Vaginismus, 205
    forcible dilatation for, 211
    Sims's operation for, 213
Vagina, atresia of, 204
    congenital absence of, 206
    non-retaining, 206
    irritable tumor of, 210
Vaginitis, 214
Vesical causes of sterility, 223
    catarrh, 223
    salicylic acid in, 225
Vesico-vaginal fistula, 229
Venery, excessive, 252

White's hysterotome, 104

www.ingramcontent.com/pod-product-compliance
Lightning Source LLC
Chambersburg PA
CBHW020307240426
43673CB00039B/729